"Being good at science and mathematics isn't just something you *are*; it's something you *become*. This users' guide to the brain unmasks the mystery around achieving success in mathematics and science. I have seen far too many students opt out when they hit a rough patch. But now that learners have a handy guide for 'knowing better,' they will also be able to 'do better.'"

—Shirley Malcom, head of education and human resources programs, American Association for the Advancement of Science

"*A Mind for Numbers* is an excellent book about how to approach mathematics, science, or any realm where problem solving plays a prominent role."

—J. Michael Shaughnessy, past president of the National Council of Teachers of Mathematics

"I have not been this excited about a book in a long time. Giving students deep knowledge on *how to learn* will lead to higher retention and student success in every field. It is a gift that will last them a lifetime."

—Robert R. Gamache, Ph.D., Associate Vice President, Academic Affairs, Student Affairs, and International Relations, University of Massachusetts, Lowell

"*A Mind for Numbers* helps put students in the driver's seat—empowering them to learn more deeply and easily. This outstanding book is also a useful resource for instructional leaders. Given the urgent need for America to improve its science and math education so it can stay competitive, *A Mind for Numbers* is a welcome find."

—Geoffrey Canada, president, Harlem Children's Zone

"An ingeniously accessible introduction to the science of human cognition—along with practical advice on how to think better."

—James Taranto, *The Wall Street Journal*

"It's easy to say 'work smarter, not harder,' but Barbara Oakley actually shows you how to do just that, in a fast-paced and accessible book that collects tips based on experience and sound science. In fact, I'm going to incorporate some of these tips into my own teaching."

—Glenn Harlan Reynolds, Beauchamp Brogan Distinguished Professor of Law, The University of Tennessee

"*A Mind for Numbers* is a splendid resource for how to approach mathematics learning and, in fact, learning in any area. Barbara Oakley's authoritative guide is based on the latest research in the cognitive sciences, and provides a clear, concise, and entertaining road map for how to get the most out of learning. This is a must-read for anyone who has struggled with mathematics and anyone interested in enhancing their learning experience."

—DAVID C. GEARY, CURATORS' PROFESSOR OF PSYCHOLOGICAL SCIENCES AND INTERDISCIPLINARY NEUROSCIENCE, UNIVERSITY OF MISSOURI

"For students afraid of math and science and for those who love the subjects, this engaging book provides guidance in establishing study habits that take advantage of how the brain works."

—DEBORAH SCHIFTER, PRINCIPAL RESEARCH SCIENTIST, SCIENCE AND MATHEMATICS PROGRAMS, EDUCATION DEVELOPMENT CENTER, INC.

a mind for
numbers

$$\left\{ \begin{array}{c} \text{a mind for} \\ \text{numbers} \end{array} \right\}$$

How to Excel at Math and Science
(Even If You Flunked Algebra)

BARBARA OAKLEY, PH.D.

A TarcherPerigee Book

tarcherperigee

An Imprint of Penguin Random House LLC
375 Hudson Street
New York, New York 10014

Most TarcherPerigee books are available at special quantity
discounts for bulk purchase for sales promotions, premiums,
fund-raising, and educational needs. Special books or book excerpts
also can be created to fit specific needs. For details, write:
SpecialMarkets@penguinrandomhouse.com.

Library of Congress Cataloging-in-Publication Data

Oakley, Barbara A.
A mind for numbers : how to excel at math and science (even if
you flunked algebra) / Barbara Oakley, Ph.D.
             p.   cm.
Includes bibliographical referecnces and index.
ISBN 978-0-399-16524-5
1. Math anxiety.   2. Mathematics—Study and teaching—
Psychological aspects.   3. Educational psychology.   I. Title.
QA11.2.O33 2014   2014003665
           501'.9—dc23

Printed in the United States of America
20   19   18   17   16   15   14   13   12

BOOK DESIGN BY ELLEN CIPRIANO

A Mind for Numbers is dedicated to Dr. Richard Felder, whose brilliance and passion have launched extraordinary improvements worldwide in the teaching of science, mathematics, engineering, and technology. My own successes, like those of tens of thousands of other educators, grow out of his fertile educational approaches. *Il miglior maestro.*

The Law of Serendipity: Lady Luck favors the one who tries

contents

foreword

Your brain has amazing abilities, but it did not come with an instruction manual. You'll find that manual in *A Mind for Numbers*. Whether you're a novice or an expert, you will find great new ways to improve your skills and techniques for learning, especially related to math and science.

Henri Poincaré was a nineteenth-century mathematician who once described how he cracked a difficult mathematical problem that he had been intensively working on for weeks without success. He took a vacation. As he was getting on a bus in the south of France, the answer to the problem suddenly came to him, unbidden, from a part of his brain that had continued to work on the problem while he was enjoying his vacation. He knew he had the right solution even though he did not write down the details until he later returned to Paris.

What worked for Poincaré can work for you too, as Barbara Oakley explains in this insightful book. Surprisingly, your brain can also work on a problem even while you are sleeping and are not aware of anything. But it does this only if you concentrate on trying to solve the problem before falling asleep. In the morning, as often as not, a fresh insight will pop to mind that can help you solve the problem. The intense effort before a vacation or falling asleep is important for priming your brain; otherwise it will work on some other problem. There is nothing special about math or science in this regard—your brain will work just as hard at solving social problems as on math and science problems, if that is what has been on your mind recently.

You will find many more insights and techniques about how to learn effectively in this fascinating and timely book, which looks at learning as an adventure rather than hard labor. You will see how you can fool yourself about whether you actually know the material; you will find ways to hold your focus and space out your practice; and you will learn to condense key ideas so you can hold them more easily in your mind. Master the simple, practical approaches outlined here and you will be able to learn more effectively and with less frustration. This wonderful guide will enrich both your learning and your life.

—Terrence J. Sejnowski, Francis Crick Professor,
Salk Institute for Biological Studies

preface

This book can make a profound difference in how you look at and understand learning. You will learn the *simplest, most effective,* and *most efficient* techniques researchers know about how to learn. And you'll have fun while you're doing it.

What's surprising is that a lot of learners use ineffective and inefficient strategies. In my laboratory, for example, we have surveyed college students about their learning. They most commonly use the strategy of *repeated reading*—simply reading through books or notes over and over. We and other researchers have found that this passive and shallow strategy often produces minimal or no learning. We call this "labor in vain"—students are putting in labor but not getting anywhere.

We don't engage in passive rereading because we are dumb or lazy. We do it because we fall prey to a cognitive illusion. When we read material over and over, the material becomes familiar and fluent, meaning it is easy for our minds to process. We then think that this easy processing is a sign that we have learned something well, even though we have not.

This book will introduce you to this and other illusions of learning and give you tools to overcome them. And it will introduce great new tools, such as retrieval practice, that can have a powerful effect in boosting the "bang for your buck" from your time spent in learning. It's a deeply practical yet inspiring book that helps you see clearly why some approaches are so much more effective than others.

We're on the edge of an explosion in knowledge about how to learn most effectively. In this new world of insight, you'll find *A Mind for Numbers* to be an indispensable guide.

—*Jeffrey D. Karpicke, James V. Bradley Associate Professor of Psychological Sciences, Purdue University*

note to the reader

People who work professionally with math and science often spend years discovering effective learning techniques. Once they've figured these methods out, *great!* They have unwittingly passed the initiation rites needed to join the mysterious society of math and science practitioners.

I've written this book to lay out these simple techniques so that you can immediately begin using them. What takes years for practitioners to discover is now at your fingertips.

Using these approaches, no matter what your skill levels in math and science, you can change your thinking and change your life. If you're already an expert, this peek under the mental hood will give you ideas for turbocharging successful learning, including counterintuitive test-taking tips and insights that will help you make the best use of your time on homework and problem sets. If you're struggling, you'll see a structured treasure trove of practical techniques that walk you through what you need to do to get on track. If you've ever wanted to become better at anything, this book will help serve as your guide.

This book is for high school students who love art and English classes but loathe math. It is meant as well as for college students who already excel in math, science, engineering, and business, but who suspect there are mental tools to be added to their learning toolkits. It's for parents whose children are either falling off the math track or trying to rocket toward math and science stardom. It's for the frazzled nine-to-five worker who hasn't been able to pass an important certification test, and for the night-shift convenience store clerk who has dreamed of becoming a nurse—or even a doctor. It's for the growing army of homeschoolers. It's for teachers and professors—not only in math, science, engineering, and technology, but also in fields such as education, psychology, and business. It's for the retiree who finally has the time to embrace new knowledge in computing, for example, or the intricacies of great cooking. And it's for readers of all ages who love to learn a little about everything.

In short, this book is for you. Enjoy!

—Barbara Oakley, Ph.D., P.E., Fellow, American Institute
for Medical & Biological Engineering and Vice President,
Institute for Electrical and Electronics Engineers—
Engineering in Medicine and Biology Society

{ 1 }

open the door

What are the odds you'd open your refrigerator door and find a zombie in there, knitting socks? The odds are about the same that a touchy-feely, language-oriented person like me would end up as a professor of engineering.

Growing up, I *hated* math and science. I flunked my way through high school math and science courses, and only started studying trigonometry—remedial trigonometry—when I was twenty-six years old.

As a youngster, even the simple concept of reading a clock face didn't seem to make sense to me. Why should the little hand point toward the hour? Shouldn't it be the big hand, since the hour was more important than the minute? Did the clock read ten ten? Or one fifty? I was perpetually confused. Worse than my problems with clocks was the television. In those days before the remote control, I didn't even know which button turned the television on. I watched a show only in the company of my brother or sister. They not only could turn the TV on, but could also tune the channel to the program we wanted to watch. Nice.

All I could conclude, looking at my technical ineptitude and flunking grades in math and science, was that I wasn't very smart. At least, not that way. I didn't realize it then, but my self-portrait as being technically, scientifically, and mathematically incapable was shaping my life. At the root of it all was my problem with mathematics. I had come to think of numbers and equations as akin to one of life's deadly diseases—to be avoided at all costs. I didn't realize then that there were simple mental tricks that could have brought math into focus for me, tricks that are helpful not only for people who are bad at math, but also for those who are already good at it. I didn't understand that my type of thinking is typical of people who believe they can't do math and science. Now, I realize that my problem was rooted in two distinctly different modes for viewing the world. Back then, I only knew how to tap one mode for learning—and the result was that I was deaf to the music of math.

Mathematics, as it's generally taught in American school systems, can be a saintly mother of a subject. It climbs logically and majestically from addition through subtraction, multiplication, and division. Then it sweeps up toward the heavens of mathematical beauty. But math can also be a wicked stepmother. She is utterly unforgiving if you happen to miss any step of the logical sequence—and missing a step is easy to do. All you need is a disruptive family life, a burned-out teacher, or an unlucky extended bout with illness—even a week or two at a critical time can throw you off your game.

Or, as was the case with me, simply no interest or seeming talent whatsoever.

In seventh grade, disaster struck my family. My father lost his job after a serious back injury. We ended up in a hardscrabble school district where a crotchety math teacher made us sit for hours in the sweltering heat doing rote addition and multiplication. It didn't help that Mr. Crotchety refused to provide any explanations. He seemed to enjoy seeing us flounder.

By this time, I not only didn't see any use for math—I actively loathed it. And as far as the sciences went—well, they didn't. In my first chemistry experiment, my teacher chose to give my lab partner

Me at age ten with Earl the lamb. I loved critters, reading, and dreaming. Math and science weren't on my play list.

and me a different substance than the rest of the class. He ridiculed us when we fudged the data in an attempt to match everyone else's results. When my well-meaning parents saw my failing grades and urged me to get help during the teacher's office hours, I felt I knew

better. Math and science were worthless, anyway. The Gods of Required Coursework were determined to shove math and science
down my throat. My way of winning was to refuse to understand
anything that was taught, and to belligerently flunk every test.
There was no way to outmaneuver my strategy.

I did have other interests, though. I liked history, social studies,
culture, and especially language. Luckily, those subjects kept my
grades afloat.

Right out of high school, I enlisted in the army because they
would actually pay me to learn another language. I did so well in
studying Russian (a language I'd selected on a whim) that an ROTC
scholarship came my way. I headed off to the University of Washington to get a bachelor's degree in Slavic languages and literature,
where I graduated with honors. Russian flowed like warm syrup—
my accent was so good that I found myself on occasion mistakenly taken for a native speaker. I spent lots of time gaining this
expertise—the better I got, the more I enjoyed what I was doing.
And the more I enjoyed what I was doing, the more time I spent on
it. My success reinforced my desire to practice, and that built more
success.

But in the most unlikely situation I could have ever imagined, I
eventually found myself commissioned as a second lieutenant in the
U.S. Army Signal Corps. I was suddenly expected to become an expert in radio, cable, and telephone switching systems. What a turning point! I went from being on top of the world, an expert linguist,
in control of my destiny, to being thrown into a new technological
world where I was as stunted as a stump.

Yikes!

I was made to enroll in mathematically oriented electronics
training (I finished at the bottom of the class), and then off I went
to West Germany, where I became a pitiable communications platoon leader. I saw that the officers and enlisted members who *were*

technically competent were in demand. They were problem solvers of the first order, and their work helped everyone accomplish the mission.

I reflected on the progress of my career and realized that I'd followed my inner passions without also being open to developing new ones. As a consequence, I'd inadvertently pigeonholed myself. If I stayed in the army, my poor technical know-how would always leave me a second-class citizen.

On the other hand, if I left the service, what could I do with a degree in Slavic languages and literature? There aren't a lot of jobs for Russian linguists. Basically, I'd be competing for entry-level secretarial-type jobs with millions of others who also had bachelor's of arts degrees. A purist might argue that I'd distinguished myself in both my studies and my service and could find much better work, but that purist would be unaware of how tough the job market can sometimes be.

Fortunately there was another unusual option. One of the great benefits of my service was that I had GI Bill money to offset the costs of future schooling. What if I used that support to do the unthinkable and try to retrain myself? Could I retool my brain from math-phobe to math lover? From technophobe to technogeek?

I'd never heard of anyone doing anything like that before, and certainly not coming from the phobic depths I'd sunk to. There couldn't possibly be anything more foreign to my personality than mastering math and science. But my colleagues in the service had shown me the concrete benefits of doing so.

It became a challenge—an irresistible challenge.

I decided to retrain my brain.

It wasn't easy. The first semesters were filled with frightening frustration. I felt like I was wearing a blindfold. The younger students around me mostly seemed to have a natural knack for seeing the solutions, while I was stumbling into walls.

But I began to catch on. Part of my original problem, I found, was that I had been putting my effort forth in the wrong way—like trying to lift a piece of lumber when you're standing on it. I began to pick up little tricks about not only how to study but when to quit. I learned that internalizing certain concepts and techniques could be a powerful tool. I also learned not to take on too much at once, allowing myself plenty of time to practice even if it meant my class-mates would sometimes graduate ahead of me because I wasn't taking as many courses each semester as they were.

As I gradually *learned how to learn* math and science, things be-came easier. Surprisingly, just as with studying language, the better I got, the more I enjoyed what I was doing. This former Queen of the Confused in math went on to earn a bachelor's degree in electri-cal engineering and then a master's in electrical and computer en-gineering. Finally, I earned a doctorate in systems engineering, with a broad background that included thermodynamics, electromag-netics, acoustics, and physical chemistry. The higher I went, the bet-ter I did. By the time I reached my doctoral studies, I was breezing by with perfect grades. (Well, perhaps not quite breezing. Good grades still took work. But the work I needed to do was clear.)

Now as a professor of engineering, I have become interested in the inner workings of the brain. My interest grew naturally from the fact that engineering lies at the heart of the medical images that allow us to tease out how the brain functions. I can now more clearly see how and why I was able to change my brain. I also see how I can help *you* learn more effectively without the frustration and struggle I experienced.[1] And as a researcher whose work straddles engineer-ing, the social sciences, and the humanities, I'm also aware of the essential creativity underlying not just art and literature, but also math and science.

If you don't (yet) consider yourself naturally good at math and science, you may be surprised to learn that **the brain is *designed* to**

do extraordinary mental calculations. We do them every time we catch a ball, or rock our body to the beat of a song, or maneuver our car around a pothole in the road. We often do complex calculations, solving complex equations unconsciously, unaware that we sometimes already know the solution as we slowly work toward it.[2] In fact, we all have a natural feel and flair for math and science. Basically, we just need to master the lingo and culture.

In writing this book, I connected with hundreds of the world's leading professor-teachers of mathematics, physics, chemistry, biology, and engineering, as well as education, psychology, neuroscience, and professional disciplines such as business and the health sciences. It was startling to hear how often these world-class experts had used precisely the approaches outlined in the book when they themselves were learning their disciplines. These techniques were also what the experts asked their students to use—but since the methods sometimes seem counterintuitive, and even irrational, instructors have often found it hard to convey their simple essence. In fact, because some of these learning and teaching methods are derided by ordinary instructors, superstar teachers sometimes divulged their teaching and learning secrets to me with embarrassment, unaware that many other top instructors shared similar approaches. By collecting many of these rewarding insights in one place, you too can easily learn and apply practical techniques gleaned in part from these "best of the best" teachers and professors. These techniques are especially valuable for helping you learn more deeply and effectively in limited time frames. You'll also gain insight from students and other fellow learners—people who share your constraints and considerations.

Remember, this is a book for math experts and mathphobes alike. This book was written to make it easier for you to learn math and science, regardless of your past grades in those subjects or how good or bad you think you are at them. It is designed to expose your thought processes so you can understand how your mind learns—and also how your mind sometimes fools you into believing you're learning, when you're actually not. The book also includes plenty of skill-building exercises that you can apply directly to your current studies. **If you're *already* good at numbers or science, the insights in this book can help make you better.** They will broaden your enjoyment, creativity, and equation-solving elegance.

If you're simply convinced you don't have a knack for numbers or science, this book may change your mind. You may find it hard to believe, but there's hope. When you follow these concrete tips based on how we actually learn, you'll be amazed to see the changes within yourself, changes that can allow new passions to bloom.

What you discover will help you be more effective and creative, not only in math and science, but in almost everything you do.

Let's begin!

{ 2 }
easy does it:

*Why Trying Too Hard Can Sometimes
Be Part of the Problem*

f you want to understand some of the most important secrets to learning math and science, look at the following picture.

The man on the right is legendary chess grand master Garry Kasparov. The boy on the left is thirteen-year-old Magnus Carlsen. Carlsen has just wandered away from the board during the height of a speed chess game, where little time is given to think about moves or strategy. That's a little like casually deciding to do a backflip while walking a tightrope across Niagara Falls.

Yes, Carlsen was psyching out his opponent. Rather than obliterating the upstart youngster, the flustered Kasparov played to a draw. But the brilliant Carlsen, who went on to become the youngest top-rated chess player in history, was doing something far beyond playing mind games with his older opponent. Gaining insight into Carlsen's approach can help us understand how the mind learns

Thirteen-year-old Magnus Carlsen (*left*), and legendary genius Garry Kasparov playing speed chess at the "Reykjavík Rapid" in 2004. Kasparov's shock is just beginning to become apparent.

math and science. Before we go into how Carlsen psyched out Kasparov, we need to cover a couple of important ideas about how people think. (But I promise, we'll come back to Carlsen.)

We're going to be touching on some of the main themes of the book in this chapter, so don't be surprised if you have to toggle around a bit in your thinking. Being able to toggle your thinking—getting a glimpse of what you are learning before returning later to more fully understand what's going on, is itself one of the main ideas in the book!

NOW YOU TRY!

Prime Your Mental Pump

As you first begin looking at a chapter or section of a book that teaches concepts of math or science, it helps to take a "picture walk" through the chapter, glancing not only at the graphics, diagrams, and photos, but also at the section headings, summary, and even questions at the end of the chapter, if the book has them. This seems counterintuitive—you haven't actually read the chapter yet, but it helps prime your mental pump. So go ahead now and glance through this chapter and th e questions at the end of the chapter.

You'll be surprised at how *spending a minute or two glancing ahead before you read in depth will help you organize your thoughts.* You're creating little neural hooks to hang your thinking on, making it easier to grasp the concepts.

Focused versus Diffuse Thinking

Since the very beginning of the twenty-first century, neuroscientists have been making profound advances in understanding the two different types of networks that the brain switches between—*highly attentive states* and more relaxed *resting state networks*.[1] We'll call the thinking processes related to these two different types of networks the **focused mode** and **diffuse mode**, respectively—these modes are highly important for learning.[2] It seems you frequently switch back and forth between these two modes in your day-to-day activities. You're in either one mode or the other—not consciously in both at the same time. The diffuse mode does seem to be able to work quietly in the background on something you are not actively focusing on.[3] Sometimes you may also flicker for a rapid moment to diffuse-mode thinking.

Focused-mode thinking is essential for studying math and science. It involves a direct approach to solving problems using rational, sequential, analytical approaches. The focused mode is associated with the concentrating abilities of the brain's prefrontal cortex, located right behind your forehead.[4] Turn your attention to something and *bam*—the focused mode is *on*, like the tight, penetrating beam of a flashlight.

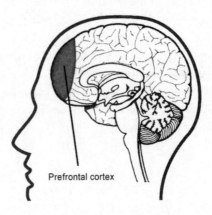

Prefrontal cortex

The prefrontal cortex is the area
right behind the forehead.

Diffuse-mode thinking is also essential for learning math and science. It allows us to suddenly gain a new insight on a problem we've been struggling with and is associated with "big-picture" perspectives. Diffuse-mode thinking is what happens when you relax your attention and just let your mind wander. This relaxation can allow different areas of the brain to hook up and return valuable insights. Unlike the focused mode, the diffuse mode seems less affiliated with any one area of the brain—you can think of it as being "diffused" throughout the brain.[5] Diffuse-mode insights often flow from preliminary thinking that's been done in the focused mode. (The diffuse mode must have clay to make bricks!)

Learning involves a complex flickering of neural processing among different areas of the brain, as well as back and forth between hemispheres.[6] So this means that thinking and learning is more complicated than simply switching between the focused and diffuse modes. But fortunately, we don't need to go deeper into the physical mechanisms. We're going to take a different approach.

The Focused Mode—A Tight Pinball Machine

To understand focused and diffuse mental processes, we're going to play some pinball. (Metaphors are powerful tools for learning in math and science.) In the old game of pinball, you pull back on a spring-loaded plunger and it whacks a ball, which ends up bouncing randomly around the circular rubber bumpers.

This happy zombie is playing neural pinball.

Look at the following illustration. When you focus your attention on a problem, your mind pulls back the mental plunger and releases a thought. Boom—that thought takes off, bumping around like the pinball in the head on the left. This is the *focused mode* of thinking.

Notice how the round bumpers are very close together in the focused mode. In contrast, the diffuse mode on the right has its circular rubber bumpers farther apart. (If you want to pursue the metaphor still further, you can think of each bumper as a cluster of neurons.)

The close bumpers of the focused mode mean that you can more easily think a precise thought. Basically, the focused mode is used to concentrate on something that's already tightly connected in your mind, often because you are familiar and comfortable with the underlying concepts. If you look closely at the upper part of the focused-mode thought pattern, you'll see a wider, "well-trodden" part of the line. That broader path shows how the focused-mode thought is following along a route you've already practiced or experienced.

For example, you can use the focused mode to multiply numbers—if you already know how to multiply, that is. If you're studying a language, you might use the focused mode to become more fluent with the Spanish verb conjugation you learned last week. If you're a swimmer, you might use the focused mode to analyze your breaststroke as you practice staying low to allow more energy to go into your forward motion.

When you focus on something, the consciously attentive prefrontal cortex automatically sends out signals along neural pathways. These signals link different areas of your brain related to what you're thinking about. This process is a little like an octopus that sends its tentacles to different areas of its surroundings to fiddle

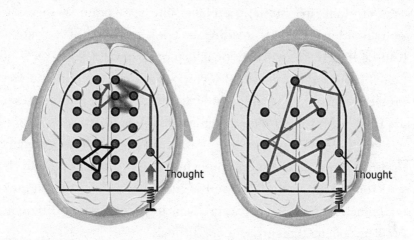

In the game "pinball," a ball, which represents a thought, shoots up from the spring-loaded plunger to bounce randomly against rows of rubber bumpers. These two pinball machines represent **focused** *(left)* and **diffuse** *(right)* ways of thinking. The focused approach relates to intense concentration on a specific problem or concept. But while in focused mode, sometimes you inadvertently find yourself focusing intently and trying to solve a problem using erroneous thoughts that are in a different place in the brain from the "solution" thoughts you need to actually need to solve the problem.

As an example of this, note the upper "thought" that your pinball first bounces around in on the left-hand image. It is very far away and completely unconnected from the lower pattern of thought in the same brain. You can see how part of the upper thought seems to have an underlying broad path. This is because you've thought something similar to that thought before. The lower thought is a new thought—it doesn't have that underlying broad pattern.

The diffuse approach on the right often involves a big-picture perspective. This thinking mode is useful when you are learning something new. As you can see, the diffuse mode doesn't allow you to focus tightly and intently to solve a specific problem—but it can allow you to get closer to where that solution lies because you're able to travel much farther before running into another bumper.

with whatever it's working on. The octopus has only so many tentacles to make connections, just as your working memory has only so many things it can hold at once. (We'll talk more about the working memory later.)

You often first funnel a problem into your brain by focusing your attention on words—reading the book or looking at your notes from a lecture. Your attentional octopus activates your focused mode. As you do your initial focused noodling around with the problem, you are thinking tightly, using the pinball bumpers that are close together to follow along familiar neural pathways related to something you already know or are familiar with. Your thoughts rattle easily through the previously ingrained patterns and quickly settle on a solution. In math and science, however, it often doesn't take much of a change for a problem to become quite different. Problem solving then grows more difficult.

Why Math and Science Can Be More Challenging

Focused problem solving in math and science is often more effort-ful than focused-mode thinking involving language and people.[7] This may be because humans haven't evolved over the millennia to manipulate mathematical ideas, which are frequently more ab-stractly encrypted than those of conventional language.[8] Obviously, we can still think *about* math and science—it's just that the *abstract-ness* and *encryptedness* adds a level—sometimes a number of levels—of complexity.

What do I mean by abstractness? You can point to a real live *cow* chewing its cud in a pasture and equate it with the letters *c-o-w* on the page. But you can't point to a real live *plus sign* that the symbol "+" is modeled after—the idea underlying the plus sign is more *ab-stract*. By *encryptedness*, I mean that one symbol can stand for a num-ber of different operations or ideas, just as the multiplication sign symbolizes repeated addition. In our pinball analogy, it's as if the abstractness and encryptedness of math can make the pinball bumpers a bit spongier—it takes extra practice for the bumpers to harden and the pinball to bounce properly. This is why dealing with procrastination, while important in studying any discipline, is particularly important in math and science. We'll be talking more about this later.

Related to these difficulties in math and science is another challenge. It's called the ***Einstellung* effect** (pronounced *EYE-nshtellung*). In this phenomenon, an idea you already have in mind, or your simple initial thought, prevents a better idea or solution from being found.[9] We saw this in the focused pinball picture, where your initial pinball thought went to the upper part of the brain, but the solution thought pattern was in the lower part of the image. (The German word *Einstellung* means "mindset"—basically you can remember *Einstellung* as installing a roadblock because of the way you are initially looking at something.)

This kind of wrong approach is especially easy to do in science because sometimes your initial intuition about what's happening is misleading. You have to unlearn your erroneous older ideas even while you're learning new ones.[10]

The *Einstellung* effect is a frequent stumbling block for students. It's not just that sometimes your natural intuitions need to be retrained—it's that sometimes it is tough even figuring out where to begin, as when tackling a homework problem. You bumble about—your thoughts far from the actual solution—because the crowded bumpers of the focused mode prevent you from springing to a new place where the solution might be found.

This is precisely why **one significant mistake students sometimes make in learning math and science is *jumping into the water before they learn to swim.***[11] In other words, they blindly start working on homework without reading the textbook, attending lectures, viewing online lessons, or speaking with someone knowledgeable.

This is a recipe for sinking. It's like randomly allowing a thought to pop off in the focused-mode pinball machine without paying any real attention to where the solution truly lies.

Understanding how to obtain *real* solutions is important, not only in math and science problem solving, but for life in general. For example, a little research, self-awareness, and even self-experimentation can prevent you from being parted with your money—or even your good health—on products that come with bogus "scientific" claims.[12] And just having a little knowledge of the relevant math can help prevent you from defaulting on your mortgage—a situation that can have a major negative impact on your life.[13]

The Diffuse Mode—A Spread-Out Pinball Machine

Think back several pages to the illustration of the diffuse-mode pinball machine brain, where the bumpers were spread far apart. This mode of thinking allows the brain to look at the world from a much broader perspective. Can you see how a thought can travel much further before it runs into a bumper? The connections are further apart—you can quickly zoom from one clump of thought to another that's quite far away. (Of course, it's hard to think precise, complex thoughts while in this mode.)

If you are grappling with a new concept or trying to solve a new problem, you don't have preexisting neural patterns to help guide your thoughts—there's no fuzzy underlying pathway to help guide you. You may need to range widely to encounter a potential solution. For this, diffuse mode is just the ticket!

Another way to think of the difference between focused and diffuse modes is to think of a flashlight. You can set a flashlight so it has a tightly focused beam that can penetrate deeply into a small area. Or you can set the flashlight onto a more diffuse setting where it casts its light broadly, but not very strongly in any one area.

If you are trying to understand or figure out something new, your best bet is to turn off your precision-focused thinking and turn on your "big picture" diffuse mode, long enough to be able to latch on to a new, more fruitful approach. As we'll see, the diffuse mode has a mind of its own—you can't simply command it to turn on. But we'll soon get to some tricks that can help you transition between modes.

COUNTERINTUITIVE CREATIVITY

"When I was learning about the diffuse mode, I began to notice it in my daily life. For instance, I realized my best guitar riffs always came to me when I was 'just messing around' as opposed to when I sat down intent on creating a musical masterpiece (in which case my songs were often clichéd and uninspiring). Similar things happened when I was writing a school paper, trying to come up with an idea for a school project, or trying to solve a difficult math problem. I now follow the rule of thumb that is basically: The harder you push your brain to come up with something creative, the less creative your ideas will be. So far, I have not found a single situation where this does not apply. Ultimately, this means that relaxation is an important part of hard work—and good work, for that matter."

—*Shaun Wassell, freshman, computer engineering*

Why Are There Two Modes of Thinking?

Why do we have these two different thinking modes? The answer may be related to two major problems that vertebrates have had in staying alive and passing their genes on to their offspring. A bird, for example, needs to focus carefully so it can pick up tiny pieces of grain as it pecks the ground for food, and at the same time, it must scan the horizon for predators such as hawks. What's the best way to carry out those two very different tasks? Split things up, of course. You can have one hemisphere of the brain more oriented toward the focused attention needed to peck at food and the other oriented toward scanning the horizon for danger. When each hemisphere tends toward a particular type of perception, it may increase the chance of survival.[14] If you watch birds, they'll first peck, and then pause to scan the horizon—almost as if they are alternating between focused and diffuse modes.

In humans, we see a similar splitting of brain functions. The left side of the brain is somewhat more associated with careful, focused attention. It also seems more specialized for handling sequential information and logical thinking—the first step leads to the second step, and so on. The right seems more tied to diffuse scanning of the environment and interacting with other people, and seems more associated with processing emotions.[15] It also is linked with handling simultaneous, big-picture processing.[16]

The slight differences in the hemispheres give us a sense of why two different processing modes may have arisen. But be wary of the idea that some people are "left-brain" or "right-brain" dominant—research indicates that is simply not true.[17] Instead it is clear that *both* hemispheres are involved in focused as well as diffuse modes of thinking. **To learn about and be creative in math and science,**

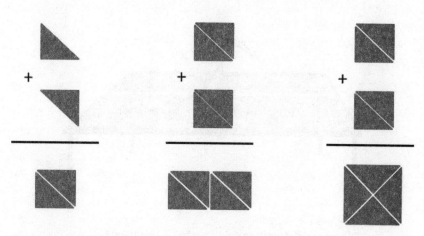

Here's a quick example that gives a sense of the difference between focused and diffuse thinking. If you are given two triangles to put together into a square shape, it's easy to do, as shown on the left. If you are given two more triangles and told to form a square, your first tendency is to erroneously put them together to form a rectangle, as shown in the middle. This is because you've already laid down a focused-mode pattern that you have a tendency to follow. It takes an intuitive, diffuse leap to realize that you need to completely rearrange the pieces if you want to form another square, as shown on the right.[18]

we need to strengthen and use both the focused and diffuse modes.[19]

Evidence suggests that to grapple with a difficult problem, we must first put hard, focused-mode effort into it. (We learned that in grade school!) Here's the interesting part: The diffuse mode is *also* often an important part of problem solving, especially when the problem is difficult. *But as long as we are consciously focusing on a problem, we are blocking the diffuse mode.*

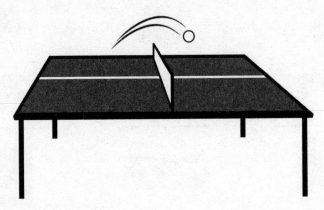

There's a winner at Ping-Pong only if the ball is able to go back and forth.

EMBRACE BEFUDDLEMENT!

....

"Befuddlement is a healthy part of the learning process. When students approach a problem and don't know how to do it, they'll often decide they're no good at the subject. Brighter students, in particular, can have difficulty in this way—their breezing through high school leaves them no reason to think that being confused is normal and necessary. But the learning process is all about working your way out of confusion. Articulating your question is 80 percent of the battle. By the time you've figured out what's confusing, you're likely to have answered the question yourself!"

—*Kenneth R. Leopold, Distinguished Teaching Professor,*
Department of Chemistry, University of Minnesota

....

The bottom line is that problem solving in any discipline often involves an exchange between the two fundamentally different modes. One mode will process the information it receives and then send the result back to the other mode. This volleying of information back and forth as the brain works its way toward a conscious solution appears essential for understanding and solving all but trivial problems and concepts.[20] The ideas presented here are extremely helpful for understanding learning in math and science. But as you are probably beginning to see, they can be just as helpful for many other subject areas, such as language, music, and creative writing.

NOW YOU TRY!

Shifting Modes

Here's a cognitive exercise that can help you feel the shift from focused to diffuse mode. See whether you can form a new triangle that points down by moving only three coins.

When you relax your mind, releasing your attention and focusing on nothing in particular, the solution can most easily come to you.

You should know that some children get this exercise instantly,

while some highly intelligent professors finally just give up. To answer this question, it helps to summon your inner child. The solutions for this challenge and for all the "Now You Try!" challenges in the book can be found in the endnotes.[21]

Procrastination Prelude

Many people struggle with procrastination. We'll have a lot to say later in this book about how to deal effectively with procrastination. For now, keep in mind that **when you procrastinate, you are leaving yourself only enough time to do superficial focused-mode learning.** You are also increasing your stress level because you know you have to complete what feels like an unpleasant task. The resulting neural patterns will be faint and fragmented and will quickly disappear—you'll be left with a shaky foundation. In math and science in particular, this can create severe problems. If you cram for a test at the last minute or quickly breeze through your homework, you won't have time for either learning mode to help you tackle the tougher concepts and problems or to help you synthesize the connections in what you are learning.

NOW YOU TRY!

Focusing Intently but Briefly

If you often find yourself procrastinating, as many of us do, here's a tip. Turn off your phone and any sounds or sights (or websites) that might signal an interruption. Then set a timer for twenty-five minutes and put yourself toward doing a twenty-five-minute interlude of work focused on a task—any task. Don't worry about finishing the task— just worry about working on it. Once the twenty-five minutes is up, reward yourself with web surfing, checking your phone, or whatever you like to do. *This reward is as important as the work itself.* You'll be amazed at how productive a focused twenty-five-minute stint can be—especially when you're just focusing on the work itself, *not* on finishing. (This method, known as the Pomodoro technique, will be discussed in more detail in chapter 6.)

If you want to apply a more advanced version of this approach, imagine that at the end of the day, you are reflecting on the *one* most important task that you accomplished that day. What would that task be? Write it down. Then work on it. Try to complete at least three of these twenty-five-minute sessions that day, on whatever task or tasks you think are most important.

At the end of your workday, look at what you crossed off your list and savor the feeling of accomplishment. Then write a few key things that you would like to work on the next day. This early preparation will help your diffuse mode begin to think about how you will get those tasks done the next day.

SUMMING IT UP

- Our brain uses two very different processes for thinking—the focused and diffuse modes. It seems you toggle back and forth between these modes, using one or the other.

- It is typical to be stumped by new concepts and problems when we first focus on them.

- To figure out new ideas and solve problems, it's important not only to focus initially, but also to subsequently turn our focus *away* from what we want to learn.

- The *Einstellung* effect refers to getting stuck in solving a problem or understanding a concept as a result of becoming fixated on a flawed approach. Switching modes from focused to diffuse can help free you from this effect. Keep in mind, then, that sometimes you will need to be flexible in your thinking. You may need to switch modes to solve a problem or understand a concept. Your initial ideas about problem solving can sometimes be very misleading.

{

PAUSE AND RECALL

Close the book and look away. What were the main
ideas of this chapter? Don't worry if you can't recall
very much when you first begin trying this. As you continue practicing this technique, you'll begin noticing
changes in how you read and how much you recall.

ENHANCE YOUR LEARNING

1. How would you recognize when you are in the diffuse
   mode? How does it feel to be in the diffuse mode?

2. When you are consciously thinking of a problem, which
   mode is active and which is blocked? What can you do to
   escape this blocking?

3. Recall an episode where you experienced the *Einstellung*
   effect. How were you able to change your thinking to get
   past the preconceived, but erroneous, notion?

4. Explain how the focused and diffuse modes might be
   equated to an adjustable beam on a flashlight. When
   can you see farther? When can you see more broadly,
   but less far?

5. Why is procrastination sometimes a special challenge for
   those who are studying math and science?

SHIFTING OUT OF BEING STUCK: INSIGHT FROM NADIA NOUI-MEHIDI, A SENIOR STUDYING ECONOMICS

"I took Calculus I in eleventh grade and it was a nightmare. It was so profoundly different from anything I had learned before that I didn't even know how to learn it. I studied longer and harder than I ever had before, yet no matter how many problems I did or how long I stayed in the library I was learning nothing. I ultimately just stuck to what I could get by with through memorizing. Needless to say, I did not do well on the AP [advanced placement] exam.

"I avoided math for the next two years, and then as a sophomore in college, I took Calculus I and got a 4.0. I don't think I was any smarter two years later, but there was a complete shift in the way I was approaching the subject.

"I think in high school I was stuck in the focused mode of thinking (*Einstellung!*) and felt that if I kept trying to approach problems in the same way it would eventually click.

"I now tutor students in math and economics and the issues are almost always that they are fixated on looking at the details of the problem for clues on how to solve it, and not on understanding the problem itself. I don't think you can tutor someone on how to think—it's kind of a personal journey. But here are some things that have helped me understand a concept that at first seems complicated or confusing.

1. I understand better when I read the book rather than listen to someone speak, so I always read the book. I skim first so I know basically what the chapter is trying to get at and then I read it in detail. I read the chapter more than once (but not in a row).

2. If after reading the book, I still don't fully understand what's going on, I Google or look at YouTube videos on the subject. This isn't because the book or professor isn't thorough, but rather because sometimes hearing a slightly different way of

phrasing something can make your mind look at the problem from a different angle and spark understanding.

3. I think most clearly when I'm driving. Sometimes I'll just take a break and drive around—this helps a lot. I have to be somewhat occupied because if I just sit down and think I end up getting bored or distracted and can't concentrate.

{ 3 }

learning is creating:

*Lessons from
Thomas Edison's Frying Pan*

Thomas Edison was one of the most prolific inventors in history, with more than one thousand patents to his name. *Nothing* got in the way of his creativity. Even as his lab was burning to the ground in a horrific accidental fire, Edison was excitedly sketching up plans for a new lab, even bigger and better than before. How could Edison be so phenomenally creative? The answer, as you'll see, relates to his unusual tricks for shifting his mode of thinking.

Shifting between the Focused and Diffuse Modes

For most people, shifting from focused to diffuse mode happens naturally if you distract yourself and then allow a little time to pass. You can go for a walk, take a nap, or go to the gym. Or you can work on something that occupies other parts of your brain: listening to music, conjugating Spanish verbs, or cleaning your gerbil cage.[1] *The key is to do something else until your brain is consciously free of any thought of the problem.* Unless other tricks are brought into play, this generally takes several hours. You may say, "I don't have that kind of time." You do, however, if you simply switch your focus to other things you need to do, and mix in a little relaxing break time.

Creativity expert Howard Gruber has suggested that one of the three *B*'s usually seems to do the trick: the bed, the bath, or the bus.[2] One remarkably inventive chemist of the mid-1800s, Alexander Williamson, observed that a solitary walk was worth a week in the laboratory in helping him progress in his work.[3] (Lucky for him there were no smartphones then.) Walking spurs creativity in many fields; a number of famous writers, such as Jane Austen, Carl Sandburg, and Charles Dickens, found inspiration during their frequent long walks.

Once you are distracted from the problem at hand, the diffuse mode has access and can begin pinging about in its big-picture way to settle on a solution.[4] After your break, when you return to the problem at hand, you will often be surprised at how easily the solution pops into place. Even if the solution doesn't appear, you will often be further along in your understanding. It can take a lot of hard focused-mode work beforehand, but the sudden, unexpected solution that emerges from the diffuse mode can make it feel almost like the "aha!" mode.

That whispered, intuitive solution to whatever puzzle you are attempting to deal with is one of the most elusively cool feelings of math and science—and art, literature, and anything else creative, for that matter! And yes, as you'll see, *math and science are deeply creative forms of thinking even when you are just learning them in school.*

That twilight, disconnected feeling one experiences while drifting off to sleep was, it seems, part of the magic behind Edison's extraordinary creativity. When faced with a difficult problem, instead of focusing intently on it, Edison, according to legend, took a

Brilliant inventor Thomas Edison (*above*) is thought to have used a clever trick to switch from focused to diffuse mode. This was the same trick used by famed surrealist painter Salvador Dalí (*below*) for his artistic creations.

nap. But he did so while sitting in a lounge chair, holding a ball bearing in his hand above a plate on the floor. As he relaxed, his thoughts moved toward free and open diffuse-mode thinking. (This is a reminder that falling asleep is a good way to get the brain thinking loosely about a problem you want to solve, or anything you are working on creatively.) When Edison fell asleep, the ball bearing fell from his hands. The clatter woke him so he could grasp the fragments of his diffuse-mode thinking to create new approaches.[5]

Creativity Is about Harnessing and Extending Your Abilities

There is a deep connection between technical, scientific, and artistic creativity. Wild surrealist painter Salvador Dalí, like Thomas Edison, also used a nap and the clatter of an object falling from his hand to tap into *his* diffuse-mode creative perspectives. (Dalí called this "sleeping without sleeping."[6]) **Enlisting the diffuse mode helps you learn at a deep and creative level.** There is much creativity underlying math and science problem solving. Many people think that there's only one way to do a problem, but there are often a number of different solutions, if you have the creativity to see them. For example, there are more than *three hundred* different known proofs of the Pythagorean theorem. As we will soon learn, technical problems and their solutions may be considered a form of poetry.

Creativity, however, is more than simply having a developed set of scientific or artistic capabilities. It is about harnessing and extending your abilities. Many people think they are not creative, when that is simply untrue. We *all* have the ability to make new neural connections and pull from memory something that was never put there in the first place—what creativity researchers Liane Gabora and Apara Ranjan refer to as "the magic of creativity."[7] Understanding how your mind works helps you better understand the creative nature of some of your thoughts.

NOW YOU TRY!

From Focused to Diffuse

Read the following sentence and identify how many errors it contains:

Thiss sentence contains threee errors.

The first two errors are easily discovered using a focused-mode approach. The third, paradoxical error becomes obvious only when you change perspectives and adopt a more diffuse approach.[8] (Remember, the solution is in the endnote.)

Working Back and Forth between Modes to Master Material

Edison's story reminds us of something else. *We learn a great deal from our failures in math and science.*[9] Know that you are making progress with each mistake you catch when trying to solve a problem—finding errors should give you a sense of satisfaction. Edison himself is said to have noted "I have not failed. I've just found 10,000 ways that won't work."[10]

Mistakes are inevitable. To work past them, start early on your assignments and, unless you are really enjoying what you are doing, *keep your working sessions short.* Remember, when you take breaks, your diffuse mode is still working away in the background. It's the best deal around—you continue to learn while you are taking it easy. Some people think they never enter diffuse mode, but that's simply not true. Every time you relax and think of nothing in particular, your brain enters into a natural default mode that's a form of diffuse thinking. Everybody does this.[11]

Sleep is probably the most effective and important factor in allowing your diffuse mode to tackle a difficult problem. But don't be fooled by the diffuse mode's seemingly easygoing, sometimes sleepy nature. One way to think of the diffuse mode is as a base station when you are mountain climbing. Base stations are essential resting spots in the long journey to difficult mountaintops. You use them to pause, reflect, check your gear, and make sure you've got the right route picked out. But you would never confuse resting at a base station with the hard work of getting to the top of a mountain. **In other words, just using your diffuse mode doesn't mean you can lollygag around and expect to get anywhere.** As the days and weeks pass, it's the distributed practice—the back and forth between focused-mode attention and diffuse-mode relaxation—that does the trick.[12]

Enlisting the focused mode, which is often what you need to do to first get a problem into your brain, requires your full attention. Studies have shown that we have only so much mental energy—willpower—for this type of thinking.[13] When your energy flags, sometimes you can take a break by jumping to other focused-type tasks, such as switching from studying math to studying French vocabulary. But the longer you spend in focused mode, the more mental resources you use. It's like a concentrated, extended set of mental weight lifting. That's why brief interludes that involve movement or talking with friends, where you don't have to concentrate intently, can be so refreshing.

You may want your learning to progress more quickly—to somehow command your diffuse mode to assimilate new ideas faster. But compare it to exercise. Constantly lifting weights won't make your muscles any bigger—your muscles need time to rest and grow before you use them again. Taking time off between weight sessions helps build strong muscles in the long run. Consistency over time is key!

USE THESE DIFFUSE-MODE TOOLS AS REWARDS
AFTER FIRM FOCUSED-MODE WORK[14]

. . . .

General Diffuse-Mode Activators
- Go to the gym
- Play a sport like soccer or basketball
- Jog, walk, or swim
- Dance
- Go for a drive (or tag along for the ride)
- Draw or paint
- Take a bath or shower
- Listen to music, especially without words
- Play songs you know well on a musical instrument
- Meditate or pray
- Sleep (the ultimate diffuse mode!)

The following diffuse-mode activators are best used briefly, as rewards. (These activities may pull you into a more focused mode than the preceding activities.)

- Play video games
- Surf the web
- Talk to friends
- Volunteer to help someone with a simple task
- Read a relaxing book
- Text friends
- Go to a movie or play
- Television (dropping a remote if you fall asleep doesn't count)

. . . .

Don't Worry about Keeping Up with the Intellectual Joneses

Students who are beginning to struggle in math and science often look at others who are intellectual racehorses and tell themselves they *have* to keep up. Then they don't give themselves the extra time they need to truly master the material, and they fall still further behind. As a result of this uncomfortable and discouraging situation, students end up unnecessarily dropping out of math and science.

Take a step back and look dispassionately at your strengths and weaknesses. If you need more time to learn math and science, that's simply the reality. If you're in high school, try to arrange your schedule to give yourself the time you need to focus on the more difficult materials, and limit these materials to manageable proportions. If you're in college, try to avoid a full load of heavy courses, especially if you are working on the side. A lighter load of math and science courses can, for many, be the equivalent of a heavy load of other types of courses. Especially in the early stages of college, avoid the temptation to keep up with your peers.

You may be surprised to discover that learning slowly can mean you learn more deeply than your fast-thinking classmates. One of the most important tricks that helped me retool my brain was learning to avoid the temptation to take too many math and science classes at once.

Avoid *Einstellung* (Getting Stuck)

Remember, accepting the first idea that comes to mind when you are working on an assignment or test problem can prevent you from finding a better solution. Chess players who experience *Ein-*

stellung truly believe they are scanning the board for a different solution. But careful study of where their eyes are moving shows that
they are keeping their focus on the original solution. *Not only their
eyes, but their mind itself can't move away enough to see a new approach to
the problem.*[15]

According to recent research, *blinking is a vital activity that provides another means of reevaluating a situation.* Closing our eyes seems
to provide a micropause that momentarily deactivates our attention
and allows us, for the briefest of moments, to refresh and renew our
consciousness and perspective.[16] So blinking may momentarily disconnect us from our focused-mode perspective. But on the other
hand, deliberately closing our eyes may help us focus more deeply—
people often look away or close or cover their eyes to avoid distractions as they concentrate on thinking of an answer.[17]

Now we can begin to understand Magnus Carlsen and his genius
for appreciating the importance of seemingly trivial distractions.
When Carlsen stood and turned his glance—and his attention—
toward other chess boards, he may have been helping his mind leap
momentarily out of focused mode. Turning his eyes and attention
elsewhere likely was critical in allowing his diffuse intuition to go to
work on his game with Kasparov. How was Carlsen able to switch
modes so quickly to gain his sudden insights? His expertise in chess
probably played a role, along with his own intuitive practice skills.
This is a hint that you, too, may be able to develop ways to quickly
toggle between the focused and diffuse modes as you develop your
expertise in a subject.

Incidentally, Carlsen probably also knew his bouncing from his
chair would distract Kasparov. Even slight distractions at that level
of play can be disconcerting—a reminder for you that deep focused attention is an important resource that you don't want to get
pulled out of. (Unless, that is, it's time to purposefully take a step
back and let the diffuse mode take over.)

Figuring out a difficult problem or learning a new concept almost always requires one or more periods when you aren't consciously working on the problem. Each interlude in which you are not directly focused on the problem allows your diffuse mode to look at it in a fresh way. When you turn your focused attention back to the problem, you consolidate new ideas and patterns that the diffuse mode has delivered.

Learning well means allowing time to pass between focused learning sessions, so the neural patterns have time to solidify properly. It's like allowing time for the mortar to dry when you are building a brick wall, as shown on the left. Trying to learn everything in a few cram sessions doesn't allow time for neural structures to become consolidated in your long-term memory—the result is a jumbled pile of bricks like those on the right.

ALTERNATING FOCUSED AND DIFFUSE THINKING

"As a piano player for a decade and a half, I sometimes found myself facing a particularly difficult run. I just couldn't get it, so I would force my fingers to do it over and over again (albeit very slowly or incorrectly), and then I'd take a break. The next day when I tried it again, I would be able to play it perfectly, as if by magic.

"I tried taking a break today with a calculus problem that was tricky and was starting to infuriate me. In the car on the way to the Renaissance festival, it came to me and I had to write it on a napkin before I forgot! (Always have napkins in your car. You never know.)"

—*Trevor Drozd, junior, computer science*

The resting times between your focused-mode efforts should be long enough to get your conscious mind completely off the problem you're working on. Usually a few hours is long enough for the diffuse mode to make significant progress but not so long that its insights fade away before being passed on to the focused mode. A good rule of thumb, when you are first learning new concepts, is not to let things go untouched for longer than a day.

The diffuse mode not only allows you to look at the material in new ways but also appears to allow you to synthesize and incorporate the new ideas in relation to what you already know. This idea of looking from fresh perspectives also gives us insight on why "sleeping on it" before making major decisions is generally a good idea,[18] and why taking vacations is important.

The tension between the focused and diffuse modes of learning takes time for your brain to resolve as you work your way through learning new concepts and solving new problems. Working in the focused mode is like providing the bricks, while working in the diffuse mode is like gradually joining the bricks together with mortar. The patient ability to keep working away, a little bit at a time, is important. This is why, if procrastination is an issue for you, it will be critical to learn some of the upcoming neural tricks to effectively address it.

NOW YOU TRY!

Observe Yourself

Next time you find yourself becoming frustrated at something or someone, try taking a mental step back and observing your reaction. Anger and frustration can occasionally have their place in motivating us to succeed, but they can also shut down key areas of the brain that we need in order to learn. Rising frustration is usually a good time-out signal for you, signaling that you need to shift to diffuse mode.

What to Do When You're *Really* Stumped

People with strong self-control can have the most difficulty in getting themselves to *turn off* their focused mode so that the diffuse mode can begin its work. After all, they've been successful because sometimes they could keep going when others flagged. If you often find yourself in this situation, you can use another trick. Make it a rule to listen to study partners, friends, or loved ones who can sense when you are becoming dangerously frustrated. Sometimes it's easier to listen to someone else than to yourself. (When my husband or children, for example, tell me to stop working with a buggy piece of software, I follow this rule myself, albeit always begrudgingly at the time.)

Speaking of talking to other people, when you're genuinely stuck, nothing is more helpful than getting insight from classmates, peers, or the instructor. Ask someone else for a different perspective on how to solve the problem or a different analogy to understand the concept; however, it's best that you first wrestle with the problem yourself *before* you talk to anyone else, because it can embed the basic concepts deeply enough that you become receptive to the explanation. Learning often means making sense of what we've ingested, and for that, we need to have ingested something. (I remember belligerently staring at my science teachers in high school, blaming them for my lack of understanding, without realizing that I was the one who needed to take the initial steps.) And don't wait until the week before midterms or final exams for this assistance. Go early and often. The teacher can often rephrase or explain in a different way that allows you to grasp the topic.

FAILURE CAN BE A GREAT TEACHER

. . . .

"When I was in tenth grade I decided to take an AP computer science class. I ended up failing the AP exam. But I would not accept the failure, so I took the class and the test again the following year. Somehow, staying away from programming for nearly a year and then coming back to it made me realize how much I truly enjoyed it. I passed the test easily on the second try. If I had been too afraid of failure to take the computer science class the first time, and then a second time, I would certainly not be what I am today, a passionate and happy computer scientist."

—*Cassandra Gordon, sophomore, computer science*

. . . .

NOW YOU TRY!

Understand the Paradoxes of Learning

Learning is often paradoxical. The very thing we need in order to learn impedes our ability to learn. We need to focus intently to be able to solve problems—yet that focus can also block us from accessing the fresh approach we may need. Success is important, but critically, so is failure. Persistence is key—but misplaced persistence causes needless frustration.

Throughout this book, you will encounter many paradoxes of learning. Can you anticipate what some of them might be?

Introduction to Working and Long-Term Memory

At this point, it's helpful to touch on some of the basics of memory. For our purposes, we're going to talk about only two major memory systems: *working memory* and *long-term memory*.[19]

Working memory is the part of memory that has to do with what you are immediately and consciously processing in your mind. It used to be thought that our working memory could hold around seven items, or "chunks," but it's now widely believed that the working memory holds only about four chunks of information. (We tend to automatically group memory items into chunks, so it seems our working memory is bigger than it actually is.[20])

You can think of working memory as the mental equivalent of a juggler. The four items only stay in the air—or in working memory—because you keep adding a little energy. This energy is needed so your metabolic vampires—natural dissipating processes—don't suck the memories away. In other words, you need to maintain these memories actively; otherwise, your body will divert your energy elsewhere, and you'll forget the information you've taken in.

Generally, you can hold about four items in your working memory, as shown in the four-item memory on the left. When you master a technique or concept in math or science, it occupies less space in your working memory. This frees your mental thinking space so that it can more easily grapple with other ideas, as shown on the right.

Your working memory is important in learning math and science because it's like your own private mental blackboard where you can jot down a few ideas that you are considering or trying to understand.

How do you keep things in working memory? Often it's through rehearsal; for example, you can repeat a phone number to yourself until you have a chance to write it down. You may find yourself shutting your eyes to keep any other items from intruding into the limited slots of your working memory as you concentrate.

In contrast, **long-term memory might be thought of as a storage warehouse.** Once items are in there, they generally stay put. The warehouse is large, with room for billions of items, and it can be easy for stored parcels to get buried so deeply that it's difficult to retrieve them. Research has shown that when your brain first puts an item of information in long-term memory, you need to revisit it a few times to increase the chances you'll later be able to find it when you need it.[21] (Techie types sometimes equate short-term memory to random-access memory [RAM], and long-term memory to hard drive space.)

Long-term memory is important for learning math and science because it is where you store the fundamental concepts and techniques that you need to use in problem solving. It takes time to move information from working memory to long-term memory. To help with this process, use a technique called *spaced repetition*. As you may have guessed, this technique involves repeating what you are trying to retain, like a new vocabulary word or a new problem-solving technique, but spacing this repetition out over a number of days.

Putting a day between bouts of repetition—extending your practice over a number of days—does make a difference. Research has shown that if you try to glue things into your memory by repeating something twenty times in one evening, for example, it won't stick nearly as well as it will if you practice it the same number of times over several days or weeks.[22] This is similar to building the brick wall we saw earlier. If you don't leave time for the mortar to dry (time for the synaptic connections to form and strengthen), you won't have a very good structure.

NOW YOU TRY!

Let Your Mind Work in the Background

The next time you are tackling a tough problem, work on it for a few minutes. When you get stuck, move on to another problem. Your diffuse mode can continue working on the tougher problem in the background. When you later return to the tougher problem, you will often be pleasantly surprised by the progress you've made.

ADVICE ON SLEEPING

"Many people will tell you that they can't nap. The one thing I learned from a single yoga class I took many years ago was to slow down my breathing. I just keep breathing slowly in and out and don't think *I must fall asleep.* Instead, I think things like, *Sleepytime!* and just focus on my breathing. I also make sure it's dark in the room, or I cover my eyes with one of those airplane sleep masks. Also, I set my phone alarm for twenty-one minutes because turning a short power nap into a longer sleep can leave you groggy. This amount of time gives me what's basically a cognitive reboot."

—*Amy Alkon, syndicated columnist and catnap queen*

The Importance of Sleep in Learning

You may be surprised to learn that simply being awake creates toxic products in your brain. During sleep, your cells shrink, causing a striking increase in the space *between* your cells. This is equivalent to turning on a faucet—it allows fluid to wash past and push the toxins out.[23] This nightly housecleaning is part of what keeps your brain healthy. When you get too little sleep, the buildup of these toxic products is believed to explain why you can't think very clearly. (Too little sleep is affiliated with conditions ranging from Alzheimer's to depression—prolonged sleeplessness is lethal.)

Studies have shown that sleep is a vital part of memory and learning.[24] Part of what this special sleep-time tidying does is erase trivial aspects of memories and simultaneously strengthen areas of importance. During sleep, your brain also rehearses some of the tougher parts of whatever you are trying to learn—going over and over neural patterns to deepen and strengthen them.[25]

Finally, sleep has been shown to make a remarkable difference in people's ability to figure out difficult problems and to find meaning and understanding in what they are learning. It's as if the complete deactivation of the conscious "you" in the prefrontal cortex helps other areas of the brain start talking more easily to one another, allowing them to put together the neural solution to your problem as you sleep.[26] (Of course, you must plant the seed for your diffuse mode by first doing focused-mode work.) It seems that if you go over the material right before taking a nap or going to sleep for the evening, you have an increased chance of dreaming about it. If you go even further and set it in mind that you *want* to dream about the material, it seems to improve your chances of dreaming about it still further.[27] Dreaming about what you are studying can substantially enhance your ability to understand—it somehow consolidates your memories into easier-to-grasp chunks.[28]

If you're tired, it's often best to just go to sleep and get up a little earlier the next day, so your reading is done with a better-rested brain. Experienced learners will attest to the fact that reading for one hour with a well-rested brain is better than reading for three hours with a tired brain. A sleep-deprived brain simply can't make the usual connections you make during normal thinking processes. Going without sleep the night before an examination can mean that even if you are perfectly prepared, your mind is simply unable to function properly, so you do poorly on the test.

A METHOD FOR MANY DISCIPLINES

Focused and diffuse approaches are valuable for all sorts of fields and disciplines, not just math and science. As Paul Schwalbe, a senior majoring in English, notes:

"If I have trouble working on a problem, I lie down in my bed with an open notebook and pen and just write out thoughts about the problem as I fall asleep, as well as sometimes right after waking up. Some of what I write makes no sense, but sometimes I gain a totally new way of looking at my problem."

SUMMING IT UP

- Use the focused mode to first start grappling with concepts and problems in math and science.
- After you've done your first hard focused work, allow the diffuse mode to take over. Relax and do something different!
- When frustration arises, it's time to switch your attention to allow the diffuse mode to begin working in the background.
- It's best to work at math and science in small doses— a little every day. This gives both the focused and diffuse modes the time they need to do their thing so you can understand what you are learning. That's how solid neural structures are built.

- If procrastination is an issue, try setting a timer for twenty-five minutes and focusing intently on your task without allowing yourself to be drawn aside by text messages, web surfing, or other attractive distractions.
- There are two major memory systems:
  - Working memory—like a juggler who can keep only four items in the air.
  - Long-term memory—like a storage warehouse that can hold large amounts of material, but needs to be revisited occasionally to keep the memories accessible.
- Spaced repetition helps move items from working memory to long-term memory.
- Sleep is a critical part of the learning process. It helps you:
  - Make the neural connections needed for normal thinking processes—which is why sleep the night before a test is so important.
  - Figure out tough problems and find *meaning* in what you are learning.
  - Strengthen and rehearse the important parts of what you are learning and prune away trivialities.

{ **PAUSE AND RECALL**

Get up and take a little break—get a glass of water or snack, or pretend you're an electron and orbit a nearby table. As you move, check your recall of the main ideas of this chapter.

ENHANCE YOUR LEARNING

1. Name some activities you would find helpful for switching from focused to diffuse mode.

2. Sometimes you can feel *certain* you have explored new approaches to analyzing a problem, when you actually haven't. What can you do to become more actively aware of your thinking processes to help keep yourself open to other possibilities? Should you *always* keep yourself open to new possibilities?

3. Why is it important to use self-control to make yourself *stop* doing something? Can you think of times outside studying and academics when this skill might also be important?

4. When you are learning new concepts, you want to review the material within a day so that the initial changes you made in your brain don't fade away. But your mind often becomes preoccupied with other matters—it's easy to let several days or more pass before you get around to looking at the material. What kind of action plan could you develop to ensure that you review important new material in a timely fashion?

NEUROPSYCHOLOGIST ROBERT BILDER'S ADVICE ON CREATIVITY

Robert Bilder *just doing it* in Makapu'u, Hawaii

Psychiatry professor Robert Bilder is the director of UCLA's Tennenbaum Center for the Biology of Creativity and leads the "Mind Well" initiative to enhance the creative achievement and psychological well-being of students, staff, and faculty at UCLA.

Research on the biology of creativity suggests several ingredients that we all can bake into our personal recipes for success. Number one is the Nike factor: Just do it!

- *Creativity is a numbers game: The best predictor of how many creative works we produce in our lifetime is . . . the number of works we produce. I sometimes find it excruciating to pull the trigger and expose my work to other people, but every time I do, it turns out for the best.*
- *Dealing with fear: A motivational poster I received after giving a talk at Facebook headquarters reads: "What would*

you do if you weren't afraid?" I try to look at this daily, and I aim to do something fearless every day. What are you afraid of? Don't let it stop you!

- *Redos come with the territory:* If you don't like the way it turned out—do it again!
- *Criticism makes us better:* By exposing our work to others, and by externalizing it so we can inspect it ourselves, we gain unique perspective and insight and develop new and improved plans for the next version.
- *Be willing to be disagreeable.* There is a negative correlation between the level of creativity and "agreeableness," so those who are the most disagreeable tend to be most creative. Looking back at the few times when I found something novel, it was because I challenged the existing answers. So I believe the creative way is advanced whenever we strip a problem back to its roots and question our own assumptions (along with assumptions suggested by others); then repeat!

{ 4 }

chunking and avoiding illusions of competence:

The Keys to Becoming an "Equation Whisperer"

Solomon Shereshevsky first came to his boss's attention because he was lazy. Or so his boss thought.

Solomon was a journalist. At that time, in the mid-1920s in the Soviet Union, being a journalist meant reporting what you were told, no more, no less. Daily assignments were given out—detailing whom to see, at what address, and to obtain what information. The editor in charge began to notice that everyone took notes. Everyone, that is, except Solomon Shereshevsky. Curious, the editor asked Solomon what was going on.

Solomon was surprised—why should he take notes, he asked, when he could remember whatever he heard? With that, Solomon repeated part of the morning's lecture, word for word. What Solomon found surprising was that he thought *everyone* had a memory like his. Perfect. Indelible.[1]

Wouldn't you love to have the gift of such a memory?

Actually, you probably wouldn't. Because hand-in-hand with his extraordinary memory, Solomon had a problem. In this chapter, we'll be talking about precisely what that problem is—involving how *focus* links to both *understanding* and *memory*.

What Happens When You Focus Your Attention?

We learned in the last chapter about that irritating situation when you become stuck in one way of looking at a problem and can't step back to see easier, better ways—*Einstellung.* Focused attention, in other words, can often help solve problems, but it can also create problems by blocking our ability to see new solutions.

When you turn your attention to something, your attentional octopus stretches its neural tentacles to connect different parts of the brain. Are you focusing on a shape? If so, one tentacle of consciousness reaches from the thalamus back toward the occipital lobe, even as another tentacle reaches toward the wrinkled surface of the cortex. The result? A whispered sense of *roundness.*

Are you focusing instead on color? The attentional tentacle in the occipital lobe shifts slightly and a sense of *green* arises.

More tentacle connections. You conclude that you are looking at a particular type of apple—a Granny Smith. Yum!

Focusing your attention to connect parts of the brain is an important part of the focused mode of learning. Interestingly, when you are stressed, your attentional octopus begins to lose the ability to make some of those connections. This is why your brain doesn't seem to work right when you're angry, stressed, or afraid.[2]

Let's say you want to learn how to speak Spanish. If you're a child hanging around a Spanish-speaking household, learning Spanish is as natural as breathing. Your mother says "mama," and you parrot "mama" back to her. Your neurons fire and wire together in a shimmering mental loop, cementing the relationship in your

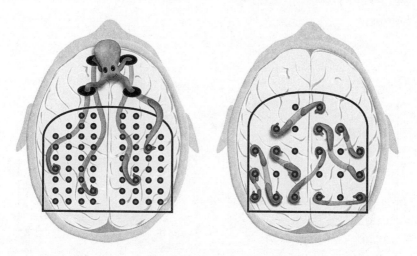

The octopus of your focused attention *(left)* reaches out through the four slots of your working memory to deliberately connect the neural bumpers of your tightly focused brain. The diffuse mode (right) has its bumpers spread farther apart. This mode consists of a wild and crazy hodgepodge of potential connections.

mind between the sound *mama* and your mother's smiling face. That scintillating neural loop is one memory trace—connected, of course, to many other related memory traces.

The best language programs—such as those at the Defense Language Institute, where I learned Russian—incorporate structured practice that includes plenty of repetition and rote, focused-mode learning of the language, along with more diffuse-like free speech with native speakers. The goal is to embed the basic words and patterns so you can speak as freely and creatively in your new language as you do in English.[3]

Focused practice and repetition—the creation of memory traces—are also at the heart of an impeccably played golf stroke, a master chef's practiced flip of an omelet, or a basketball free throw. In dance, it's a long way from a toddler's clumsy pirouette to the choreographed grace of a professional dancer. But that path to expertise is built bit by bit. Small memorized free spins, heel turns, and kicks become incorporated into larger, more creative interpretations.

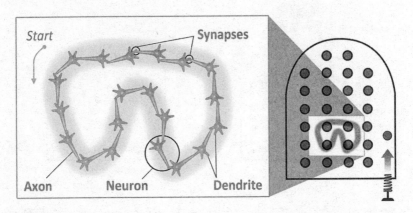

The left image symbolizes the compact connections when one chunk of knowledge is formed—neurons that fire together wire together. The image on the right shows the same pattern in your mind's symbolic pinball machine. Such a memory trace is easy to recall when you need it.

What Is a Chunk? Solomon's Chunking Problem

Solomon Shereshevsky's extraordinary memory came with a surprising drawback. His individual memory traces were each so colorful and emotional—so rich with connections—that they interfered with his ability to put those traces together and create conceptual **chunks**. He couldn't see the forest, in other words, because his imagery of each of the individual trees was so vivid.

Chunks are pieces of information that are bound together through meaning. You can take the letters *p*, *o*, and *p* and bind them into one conceptual, easy-to-remember chunk, the word *pop*. It's like converting a cumbersome computer file into a .zip file. Underneath that simple *pop* chunk is a symphony of neurons that have learned to trill in tune with one another. The complex neural activity that ties together our simplifying, abstract chunks of thought—

whether those thoughts pertain to acronyms, ideas, or concepts—
are the basis of much of science, literature, and art.

Let's take an example. In the early 1900s, German researcher
Alfred Wegener put together his theory of continental drift. As Wegener analyzed maps and thought about the information he'd
gleaned from his studies and exploration, he realized that the different land masses fit together like puzzle pieces. The similarity of
rocks and fossils between the land masses reinforced the fit. Once
Wegener put the clues together, it was clear that all the continents
had once, very long ago, been joined together in a single landmass.
Over time, the mass had broken up and the pieces had drifted apart
to form the continents separated by oceans we see today.

Continental drift! Wow—what a great discovery!

But if Solomon Shereshevsky had read this same story about the
discovery of continental drift, he wouldn't have gotten the point.
Even though he would have been able to repeat every individual
word in the story, the concept of continental drift would have been
very difficult for him to grasp, since he was unable to link his individual memory traces together to create conceptual chunks.

As it turns out, **one of the first steps toward gaining expertise
in math and science is to create conceptual chunks—mental leaps
that unite separate bits of information through meaning.**[4] Chunking the information you deal with helps your brain run more efficiently. Once you chunk an idea or concept, you don't need to
remember all the little underlying details; you've got the main
idea—the chunk—and that's enough. It's like getting dressed in the
morning. Usually you just think one simple thought—*I'll get dressed*.
But it's amazing when you realize the complex swirl of underlying
activities that take place with that one simple chunk of a thought.

When you are studying math and science, then, how do you
form a chunk?

Basic Steps to Forming a Chunk

Chunks related to different concepts and procedures can be molded in many different ways. It's often quite easy. You formed a simple chunk, for example, when you grasped the idea of continental drift. But since this is a book about how to learn math and science in general rather than geology in particular, we're going to take as our initial, illustrative chunk *the ability to understand and work a certain type of math or science problem.*

When you are learning new math and science material, you are almost always given sample problems with worked-out solutions. This is because, when you are first trying to understand how to work a problem, you have a heavy cognitive load—so it helps to start out with a fully worked-through example. It's like using a GPS unit when you are driving on unfamiliar roads in the middle of the night. Most of the details in the worked-out solution are right there, and your task is simply to figure out why the steps are taken the way they are. That can help you see the key features and underlying principles of a problem.

Some instructors do not like to give students extra worked-out problems or old tests, as they think it makes matters too easy. But there is bountiful evidence that having these kinds of resources available helps students learn much more deeply.[5] The one concern about using worked-out examples to form chunks is that it can be all too easy to focus too much on why an individual step works and not on the *connection* between steps—that is, on why this particular step is the next thing you should do. So keep in mind that I'm not talking about a cookie-cutter "just do as you're told" mindless approach when following a worked-out solution. It's more like using a guide to help you when traveling to a new place. Pay attention to

Raw information   Memorization without understanding   Information is chunked and understood

When you first look at a brand-new concept in science or math, it sometimes doesn't make much sense, as shown by the puzzle pieces above on the left. Just memorizing a fact *(center)* without understanding or context doesn't help you understand what's really going on, or how the concept fits together with the other concepts you are learning—notice there are no interlocking puzzle edges on the piece to help you fit into other pieces. **Chunking** *(right)* is the mental leap that helps you unite bits of information together through meaning. The new logical whole makes the chunk easier to remember, and also makes it easier to fit the chunk into the larger picture of what you are learning.

what's going on around you when you're with the guide, and soon you'll find yourself able to get there on your own. You will even begin to figure out new ways of getting there that the guide didn't show you.

1. **The first step in chunking, then, is to simply *focus your attention* on the information you want to chunk.**[6] If you have the television going in the background, or you're looking up every few minutes to check or answer your phone or computer messages, it means that you're going to have difficulty making a chunk, because your brain is not really focusing on the chunking. When you first begin to learn something, you are making new neural patterns and connecting them with preexisting patterns that are spread through many areas of the brain.[7] Your octopus tentacles can't make connections very well if some of them are off on other thoughts.

2. **The second step in chunking is to *understand* the basic idea you are trying to chunk,** whether it is understanding a concept such as continental drift, the idea that force is proportional to mass, the economic principle of supply and demand, or a particular type of math problem. Although this step of basic understanding—synthesizing the gist of what's important—was difficult for Solomon Shereshevsky, most students figure out these main ideas naturally. Or at least, they can grasp those ideas if they allow the focused and diffuse modes of thinking to take turns in helping them figure out what's going on.

Understanding is like a superglue that helps hold the underlying memory traces together. It creates broad, encompassing traces that link to many memory traces.[8] Can you create a chunk if you don't understand? Yes, but it's a useless chunk that won't fit in with other material you are learning.

That said, it's important to realize that *just understanding how a problem was solved does not necessarily create a chunk that you can easily call to mind later.* Do *not* confuse the "aha!" of a breakthrough in understanding with solid expertise! (That's part of why you can grasp an idea when a teacher presents it in class, but if you don't review it fairly soon after you've first learned it, it can seem incomprehensible when it comes time to prepare for a test.) Closing the book and testing yourself on how to solve the problems will also speed up your learning at this stage.

3. **The third step to chunking is gaining context so you see not just how, but also *when* to use this chunk.** Context means going beyond the initial problem and seeing more

broadly, repeating and practicing with both related and unrelated problems so you see not only when to use the chunk, but when *not* to use it. This helps you see how your newly formed chunk fits into the bigger picture. In other words, you may have a tool in your strategy or problem-solving toolbox, but if you don't know when to use that tool, it's not going to do you a lot of good. Ultimately, practice helps you broaden the networks of neurons connected to your chunk, ensuring that it is not only firm, but also accessible from many different paths.

There are chunks related to both concepts and procedures that reinforce one another. Solving a lot of math problems provides an opportunity to learn why the procedure works the way it does or why it works at all. Understanding the underlying concept makes it easier to detect errors when you make them. (Trust me, you *will* make errors, and that's a good thing.) It also makes it much easier to apply your knowledge to novel problems, a phenomenon called *transfer.* We'll talk more about transfer later.

As you can see from the following "top-down, bottom-up" illustration, learning takes place in two ways. There is a **bottom-up chunking process** where practice and repetition can help you both build and strengthen each chunk, so you can easily gain access to it when needed. And there is a **top-down "big picture" process** that allows you to see where what you are learning fits in.[9] *Both processes are vital in gaining mastery over the material.* Context is where bottom-up and top-down learning meet. To clarify here—chunking may involve your learning *how* to use a certain problem-solving technique. Context means learning *when* to use that technique instead of some other technique.

Those are the essential steps to making a chunk and fitting that chunk into a greater conceptual overview of what you are learning. But there's more.

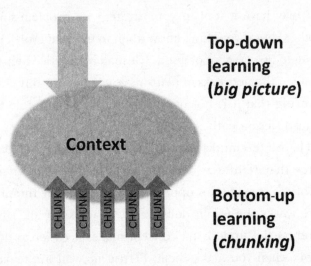

Top-down learning (*big picture*)

Context

Bottom-up learning (*chunking*)

CHUNK CHUNK CHUNK CHUNK CHUNK

Both top-down, big-picture learning, and bottom-up chunking are important in becoming an expert in math and science.

NOW I LAY ME DOWN TO SLEEP
....

"I tell my students that internalizing the accounting fundamentals is like internalizing how to type on a keyboard. In fact, as I write this myself, I'm not thinking of the act of typing, but of formulating my thoughts—the typing comes naturally. My mantra at the end of each class is to tell students to look at the Debit and Credit Rules as well as the Accounting Equation just before they tuck themselves in at night. Let those be the last things they repeat to themselves before falling asleep. Well, except meditation or prayers, of course!"

—*Debra Gassner Dragone, Accounting Instructor, University of Delaware*

....

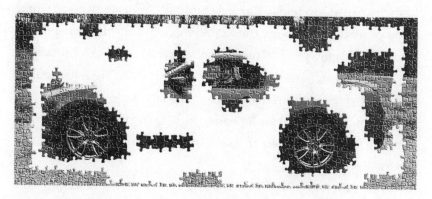

Skimming through a chapter or listening to a very well-organized lecture can allow you to gain a sense of the big picture. This can help you know where to put the chunks you are constructing. Learn the major concepts or points first—these are often the key parts of a good instructor or book chapter's outline, flow charts, tables, or concept maps. Once you have this done, fill in the details. Even if a few of the puzzle pieces are missing at the end of your studies, you can still see the big picture.

Illusions of Competence and the Importance of Recall

Attempting to *recall* the material you are trying to learn—retrieval practice—is far more effective than simply rereading the material.[10] Psychologist Jeffrey Karpicke and his colleagues have shown that many students experience *illusions of competence* when they are studying. Most students, Karpicke found, "repeatedly read their notes or textbook (despite the limited benefits of this strategy), but relatively few engage in self-testing or retrieval practice while studying."[11] When you have the book (or Google!) open right in front of you, it provides the illusion that the material is also in your brain. *But it's not.* Because it can be easier to look at the book instead of recalling, students persist in their illusion—studying in a far less productive way.

This, indeed, is why just *wanting* to learn the material, and spending a lot of time with it, doesn't guarantee you'll actually learn it. As Alan Baddeley, a renowned psychologist and expert on memory, notes: **"Intention to learn is helpful only if it leads to the use of good learning strategies."**[12]

You may be surprised to learn that highlighting and underlining must be done carefully—otherwise they can be not only ineffective but also misleading. It's as if the motion of your hand can fool you into thinking you've placed the concept in your brain. When marking up the text, train yourself to look for main ideas before making any marks, and keep your text markings to a minimum— one sentence or less per paragraph.[13] Words or notes in a margin that synthesize key concepts are a good idea.[14]

Using recall—mental retrieval of the key ideas—rather than passive rereading will make your study time more focused and effective. The only time rereading text seems to be effective is if you let time pass between rereadings so that it becomes more of an exercise in spaced repetition.[15]

Along these same lines, always work through homework problems in math and science on your own. Some textbooks include solutions at the back of the book, but you should look at these only to check your answer. This will help ensure that the material is more deeply rooted in your mind and make it much more accessible when you really need it. This is why instructors place so much emphasis on showing your work and giving your reasoning on tests and homework problems. Doing so forces you to think your way through a problem and provides a self-test of your understanding. This additional information about your thinking also gives graders a better opportunity to provide useful feedback.

You don't want to wait too long for the recall practice, so that you have to start the reinforcement of the concept from scratch every time. Try to touch again on something you're learning within a day, especially if it's new and rather challenging. This is why many professors recommend that, if at all possible, you rewrite your notes during the evening after a lecture. This helps to solidify newly forming chunks and reveals the holes in your understanding that professors just *love* to target on tests. Knowing where the holes are, of course, is the first step toward getting them filled in.

Once you've got something down, you can expand the time between "upkeep" repetitions to weeks or months—and eventually it can become close to permanent. (Returning to Russia on a visit, for example, I found myself annoyed by an unscrupulous taxi driver. To my amazement, words I hadn't thought or used for twenty-five years popped from my mouth—I hadn't even been consciously aware I *knew* those words!)

MAKE YOUR KNOWLEDGE SECOND NATURE

. . . .

"Getting a concept in class versus being able to apply it to a genuine physical problem is the difference between a simple student and a full-blown scientist or engineer. The only way I know of to make that jump is to work with the concept until it becomes second nature, so you can begin to use it like a tool."

—*Thomas Day, Professor of Audio Engineering, McNally Smith College of Music*

. . . .

Later, we'll discuss useful apps and programs that can help with learning. But for now, it's worth knowing that well-designed electronic flash card systems, such as Anki, have built into them the appropriate spaced repetition time to optimize the rate of learning new material.

One way to think about this type of learning and recall is shown in the following working-memory illustration. As we mentioned earlier, there are four or so spots in working memory.

When you are first chunking a concept, its pre-chunked parts take up all your working memory, as shown on the left. As you begin to chunk the concept, you will feel it connecting more easily and smoothly in your mind, as shown in the center. Once the concept is chunked, as shown at the right, it takes up only one slot in working memory. It simultaneously becomes one smooth strand that is easy to follow and use to make new connections. The rest of your working memory is left clear. That dangling strand of chunked material has, in some sense, increased the amount of information available to your working memory, as if the slot in working memory is a hyperlink that has been connected to a big webpage.[16]

When you are first learning how to solve a problem, your entire working memory is involved in the process, as shown by the mad tangle of connections between the four slots of working memory on the left. But once you become smoothly familiar with the concept or method you are learning and have it encapsulated as a single chunk, it's like having one smooth ribbon of thought, as shown on the right. The chunking, which enlists long-term memory, frees the rest of your working memory to process other information. Whenever you want, you can slip that ribbon (chunk) from long-term memory into your working memory and follow along the strand, smoothly making new connections.

Now you understand why it is key that you are the one doing the problem solving, not whoever wrote the solution manual. If you work a problem by just looking at the solution, and then tell yourself, "Oh yeah, I see why they did that," then the solution is not really yours—you've done almost nothing to knit the concepts into your underlying neurocircuitry. Merely glancing at the solution to a problem and thinking you truly know it yourself is one of the most common illusions of competence in learning.

NOW YOU TRY!

Understanding Illusions of Competence

Anagrams are rearrangements of letters so that one word or phrase can spell something different. Let's say you have the phrase "Me, radium ace." Can you rearrange it to spell the last name of a honorific famous physicist?[17] It may take you a bit of thought to do it. But if you saw the solution here on the page, your subsequent "aha!" feeling would make you think that your anagram-solving skills are better than they actually are.

Similarly, students often erroneously believe that they are learning by simply rereading material that is on the page in front of them. They have an illusion of competence *because the solution is already there.*[18]

Pick a mathematical or scientific concept from your notes or from a page in the book. Read it over, then look away and see what you can recall—working toward understanding what you are recalling at the same time. Then glance back, reread the concept, and try it again.

At the end of this exercise, you will probably be surprised to see how much this simple recall exercise helped improve your *understanding* of the concept.

You must have information persisting in your memory if you are to master the material well enough to do well on tests and think creatively with it.[19] The ability to combine chunks in novel ways underlies much of historical innovation. Steven Johnson, in his brilliant book *Where Good Ideas Come From*, describes the "slow hunch"—the gentle, years-long simmering of focused and diffuse processes that has resulted in creative breakthroughs ranging from Darwin's evolutionary theory to the creation of the World Wide Web.[20] Key to the slow hunch is simply having mental access to aspects of an idea. That way, some aspects can tentatively and randomly combine with others until eventually, beautiful novelty can emerge.[21] Bill Gates and other industry leaders, Johnson notes, set aside extended, weeklong reading periods so that they can hold many and varied ideas in mind during one time. This fosters their own innovative thinking by allowing fresh-in-mind, not-yet-forgotten ideas to network among themselves. (An important side note here is that a key difference between creative scientists and technically competent but nonimaginative ones is their breadth of interest.[22])

The bigger your chunked mental library, the more easily you will be able to solve problems. Also, as you gain more experience in chunking, you will see that the chunks you are able to create are bigger—the ribbons are longer.

You may think there are so many problems and concepts just in a single chapter of the science or math subject you are studying that there's no way to do them all! This is where the **Law of Serendipity** comes to play: **Lady Luck favors the one who tries.**[23]

Just focus on whatever section you are studying. You'll find that once you put the first problem or concept in your library, *whatever it is*, then the second concept will go in a bit more easily. And the third more easily still. Not that all of this is a snap, but it does get easier.

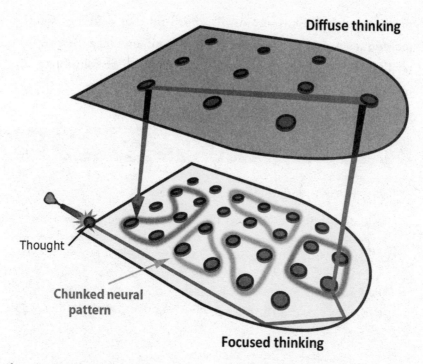

Diffuse thinking

Thought

Chunked neural pattern

Focused thinking

If you have a library of concepts and solutions internalized as chunked patterns, you can more easily skip to the right solution to a problem by listening to the whispers from your diffuse mode. Your diffuse mode can also help you connect two or more chunks together in new ways to solve unusual problems.

There are two ways to solve problems—first, through sequential, step-by-step reasoning, and second, through more holistic intuition. Sequential thinking, where each small step leads deliberately toward the solution, involves the focused mode. Intuition, on the other hand, often seems to require a creative, diffuse mode linking of several seemingly different focused mode thoughts.

Most difficult problems are solved through intuition, because they make a leap away from what you are familiar with.[24] Keep in mind that the diffuse mode's semi-random way of making connections means that the solutions it provides with should be carefully verified using the focused mode. Intuitive insights aren't always correct![25]

In building a chunked library, you are training your brain to recognize not only a specific problem, but different *types* and *classes* of problems so that you can automatically know how to quickly solve whatever you encounter. You'll start to see patterns that simplify problem solving for you and will soon find that different solution techniques are lurking at the edge of your memory. Before midterms or finals, it is easy to brush up and have these solutions at the mental ready.

NOW YOU TRY!

What to Do If You Can't Grasp It

If you don't understand a method presented in a course you are taking, stop and work backward. Go to the Internet and discover who first figured out the method or some of the earliest people to use it. Try to understand how the creative inventor arrived at the idea and why the idea is used—you can often find a simple explanation that gives a basic sense of why a method is being taught and why you would want to use it.

Practice Makes Permanent

I've already mentioned that just *understanding* what's going on is *not* usually enough to create a chunk. You can get a sense of what I mean by looking at the "brain" picture shown on p. 69. The chunks (loops) shown are really just extended memory traces that have arisen because you have knit together an understanding. A chunk, in other words, is simply a more complex memory trace. At the top is a faint chunk. That chunk is what begins to form after you've understood a concept or problem and practiced just a time or two. In the middle, the pattern is darker. This is the stronger neural pattern that results after you've practiced a little more and seen the chunk in more contexts. At the bottom, the chunk is very dark.

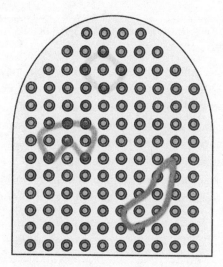

Solving problems in math and science is like playing
a piece on the piano. The more you practice,
the firmer, darker, and stronger your
mental patterns become.

You've now got a solid chunk that's firmly embedded in long-term memory.

Incidentally, strengthening an initial learning pattern within a day after you first begin forming it is important. Without the strengthening, the pattern can quickly fade away. Later, we'll talk more about the importance of spaced repetition in learning. Also, you can reinforce a "wrong" process by doing the same problems over and over the wrong way. This is why checking things is so important. Even getting the right answer can occasionally mislead you if you get it by using an incorrect procedure.

THE IMPORTANCE OF CHUNKING

"Mathematics is amazingly compressible: you may struggle a long time, step by step, to work through the same process or idea from several approaches. But once you really understand it and have the mental perspective to see it as a whole, there is often a tremendous mental compression. You can file it away, recall it quickly and completely when you need it, and use it as just one step in some other mental process. The insight that goes with this compression is one of the real joys of mathematics."[26]

—*William Thurston, winner of the Fields Medal,*
the top award in mathematics

The challenge with repetition and practice, which lie behind the mind's creation of solid chunks, is that it can be boring. Worse yet, in the hands of a poor instructor, like my old math teacher, Mr. Crotchety, practice can become an unrelenting instrument of torture. Despite its occasional misuse, however, it's critical. Everybody knows you can't effectively learn the chunked patterns of chess, language, music, dance—just about anything worthwhile—without repetition. Good instructors can explain why the practice and repetition is worth the trouble.

Ultimately, both bottom-up chunking and top-down big-picture approaches are vital if you are to become an expert with the material. We love creativity and the idea of being able to learn by seeing the big picture. **But you can't learn mathematics or science without also including a healthy dose of practice and repetition to help you build the chunks that will underpin your expertise.**[27]

Research published in the journal *Science* provided solid evidence along these lines.[28] Students studied a scientific text and then practiced it by recalling as much of the information as they could. Then they restudied the text and recalled it (that is, tried to remember the key ideas) once more.

The results?

In the same amount of time, *by simply practicing and recalling the* *material,* **students learned far more and at a much deeper level** **than they did using any other approach,** including simply rereading the text a number of times or drawing concept maps that supposedly enriched the relationships in the materials under study. This improved learning comes whether students take a formal test or just informally test themselves.

This reinforces an idea we've alluded to already. When we retrieve knowledge, we're not being mindless robots—*the retrieval pro-* *cess itself enhances deep learning and helps us begin forming chunks.*[29] Even more of a surprise to researchers was that the students themselves predicted that simply reading and recalling the materials wasn't the best way to learn. They thought concept mapping (drawing diagrams that show the relationship between concepts) would be best. But if you try to build connections between chunks *before the* *basic chunks are embedded in the brain,* it doesn't work as well. It's like trying to learn advanced strategy in chess before you even understand the basic concepts of how the pieces move.[30]

Practicing math and science problems and concepts in a variety of situations helps you build chunks—solid neural patterns with deep, contextual richness.[31] The fact is, when learning *any* new skill or discipline, you need plenty of varied practice with different contexts. This helps build the neural patterns you need to make the new skill a comfortable part of your way of thinking.

KEEP YOUR LEARNING AT THE TIP OF YOUR TONGUE
....

"By chance, I have used many of the learning techniques de-scribed in this book. As an undergraduate I took physical chem-istry and became fascinated with the derivations. I got into a habit of doing every problem in the book. As a result, I hard-wired my brain to solve problems. By the end of the semester I could look at a problem and know almost immediately how to solve it. I suggest this strategy to my science majors in particular, but also to the nonscientists. I also talk about the need to study every day, not necessarily for long periods of time but just enough to keep what you are learning at the tip of your tongue. I use the example of being bilingual. When I go to France to work, my French takes a few days to kick in, but then it is fine. When I return to the States and a student or colleague asks me some-thing on my first or second day back, I have to search for the English words! When you practice every day the information is just there—you do not have to search for it."

—Robert R. Gamache, Associate Vice President,
Academic Affairs, Student Affairs, and International
Relations, University of Massachusetts, Lowell

....

Recall Material While Outside Your Usual Place of Study: The Value of Walking

Doing something physically active is especially helpful when you have trouble grasping a key idea. As mentioned earlier, stories abound of innovative scientific breakthroughs that occurred when the people who made them were out walking.[32]

In addition, **recalling material when you are outside your usual place of study helps you strengthen your grasp of the material by viewing it from a different perspective.** People sometimes lose sub-conscious cues when they take a test in a room that looks different from where they studied. By thinking about the material while you are in various physical environments, you become independent of cues from any one location, which helps you avoid the problem of the test room being different from where you originally learned the material.[33]

Internalizing math and science concepts can be *easier* than memorizing a list of Chinese vocabulary words or guitar chords. After all, you've got the problem there to speak to you, telling you what you need to do next. In that sense, problem solving in math and science is like dance. In dance, you can *feel* your body hinting at the next move.

Different types of problems have different review time frames that are specific to your own learning speed and style.[34] And of course, you have other obligations in your life besides learning one particular topic. You have to prioritize how much you're able to do, also keeping in mind that you *must* schedule some time off to keep your diffuse mode in play. How much internalizing can you do at a stretch? It depends—everyone is different. But, here's the real beauty of internalizing problem solutions in math and science. The more you do it, the easier it becomes, and the more useful it is.

ORGANIZE, CHUNK—AND SUCCEED

"The first thing I always do with students who are struggling is ask to see how they are organizing their notes from class and reading. We often spend most of the first meeting going over ways they can organize or chunk their information rather than with my explaining concepts. I have them come back the next week with their material already organized, and they are amazed at how much more they retain."

—*Jason Dechant, Ph.D., Course Director, Health Promotion and Development, School of Nursing, University of Pittsburgh*

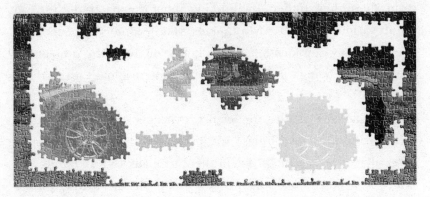

If you don't practice with your growing chunks, it is harder to put together the big picture—the pieces are simply too faint.

Interleaving—Doing a Mixture of Different Kinds of Problems—versus Overlearning

One last important tip in becoming an equation whisperer is interleaving.[35] **Interleaving means practice by doing a mixture of different kinds of problems requiring different strategies.**

When you are learning a new problem-solving approach, either from your teacher or from a book, you tend to learn the new technique and then practice it over and over again during the same study session. Continuing the study or practice after it is well understood is called *overlearning*. Overlearning can have its place—it can help produce an automaticity that is important when you are executing a serve in tennis or playing a perfect piano concerto. But be wary of repetitive overlearning during a single session in math and science learning—research has shown it can be a waste of valuable learning time.[36] (Revisiting the approach mixed with other approaches during a subsequent study session, however, is just fine.)

In summary, then, once you've got the basic idea down during a session, continuing to hammer away at it during the same session doesn't necessarily strengthen the kinds of long-term memory connections you want to have strengthened. Worse yet, focusing on one technique is a little like learning carpentry by only practicing with a hammer. After a while, you think you can fix anything by just bashing it.[37]

The reality is, mastering a new subject means learning to select and use the proper technique for a problem. The only way to learn that is by practicing with problems that require *different* techniques. Once you have the basic idea of a technique down during your study session (sort of like learning to ride a bike with training wheels), start interleaving your practice with problems of different types.[38] Sometimes this can be a little tough to do. A given section in a book, for example, is often devoted to a specific technique, so when you flip to that section, you already know which technique you're going to use.[39] Still, do what you can to mix up your learning. It can help to look ahead at the more varied problem sets that are sometimes found at the end of chapters. Or you can deliberately try to make yourself occasionally pick out why some problems call for one technique as opposed to another. **You want your brain to become used to the idea that just knowing *how* to use a particular problem-solving technique isn't enough—you also need to know *when* to use it.**

Consider creating index cards with the problem question on one side, and the question and solution steps on the other. That way you can easily shuffle the cards and be faced with a random variety of techniques you must call to mind. When you first review the cards, you can sit at a desk or table and see how much of the solution you can write on a blank sheet of paper without peeking at the back of the card. Later, when mastery is more certain, you can review your cards anywhere, even while out for a walk. Use the initial question as a cue to bring to mind the steps of the response, and flip the card over if necessary to verify that you've got the procedural steps all in mind. You are basically strengthening a new chunk. Another idea is to open the book to a randomly chosen page and work a problem while, as much as possible, hiding from view everything but the problem.

EMPHASIZE INTERLEAVING INSTEAD OF OVERLEARNING

Psychologist Doug Rohrer of the University of South Florida has done considerable research on overlearning and interleaving in math and science. He notes:

> "Many people believe overlearning means studying or practicing until mastery is achieved. However, in the research literature, overlearning refers to a learning strategy in which a student continues to study or practice immediately after some criterion has been achieved. An example might be correctly solving a certain kind of math problem and then immediately working several more problems of the same kind. Although working more problems of the same kind (rather than fewer) often boosts scores on a subsequent test, doing too many problems of the same kind in immediate succession provides diminishing returns.
>
> "In the classroom and elsewhere, students should maximize the amount they learn per unit time spent studying or practicing—that is, they should get the most bang for the buck. How can students do this? The scientific literature provides an unequivocal answer: Rather than devote a long session to the study or practice of the same skill or concept so that overlearning occurs, students should divide their effort across several shorter sessions. This doesn't mean that long study sessions are necessarily a bad idea. Long sessions are fine as long as students don't devote too much time to any one skill or concept. Once they understand 'X,' they should move on to something else and return to 'X' on another day."[40]

It's best to write the initial solution, or diagram, or concept, out by hand. There's evidence that writing by hand helps get the ideas into mind more easily than if you type the answer.[41] More than that, it's often easier to write symbolic material like Σ or Ω by hand than to search out the symbol and type it (unless you use the symbols often enough to memorize the alt codes).[42] But if you then want to photograph or scan the question and your handwritten solution to load it into a flash card program for your smartphone or laptop, that will work just fine. Beware—a common illusion of competence is to continue practicing a technique you know, simply because it's easy and it feels good to successfully solve problems. Interleaving your studies—making a point to review for a test, for example, by skipping around through problems in the different chapters and materials—can sometimes seem to make your learning more difficult. But in reality, it helps you learn more deeply.

AVOID MIMICKING SOLUTIONS—PRACTICE CHANGING MENTAL GEARS

"When students do homework assignments, they often have ten identical problems in a row. After the second or third problem, they are no longer thinking; they are mimicking what they did on the previous problem. I tell them that, when doing the homework from section 9.4, after doing a few problems, go back and do a problem from section 9.3. Do a couple more 9.4 problems, and then do one from section 9.1. This will give them practice in mentally shifting gears in the same way they'll need to switch gears on the test.

"I also believe too many students do homework just to get it done. They finish a problem, check their answer in the back of the text, smile, and go on to the next problem. I suggest that they insert a step between the smile and going on to the next problem—asking themselves this question: How would I know how to do the problem this way if I saw it on a test mixed together with other problems and I didn't know it was from this section of the text? Students need to think of every homework problem in terms of test preparation and not as part of a task they are trying to complete."

—Mike Rosenthal, Senior Instructor of Mathematics,
Florida International University

....

SUMMING IT UP

- Practice helps build strong neural patterns—that is, conceptual chunks of understanding.
- Practice gives you the mental fluidity and agility you need for tests.
- Chunks are best built with:
    - *Focused attention.*
    - *Understanding* of the basic idea.
    - *Practice* to help you gain big-picture context.
- Simple recall—trying to remember the key points without looking at the page—is one of the best ways to help the chunking process along.

In some sense, recall helps build neural hooks
that you can hang your thinking on.

ENHANCE YOUR LEARNING

1. How is a chunk related to a memory trace?

2. Think of a topic you are passionate about. Describe a chunk involving that topic that was at first difficult for you to grasp but now seems easy.

3. What is the difference between top-down and bottom-up approaches to learning? Is one approach preferable to the other?

4. Is *understanding* enough to create a chunk? Explain why or why not.

5. What is your own most common illusion of competence in learning? What strategy can you use to help avoid falling for this illusion in the future?

PAUSE AND RECALL

Next time you are with a family member, friend, or classmate, relate the essence of what you have been learning, either from this book or in regard to a class you are taking. Retelling whatever you are learning about not only helps fuel and share your own enthusiasm, but also clarifies and cements the ideas in your mind, so you'll remember them better in the weeks and months to come. Even if what you are studying is very advanced, simplifying so you can explain to others who do not share your educational background can be surprisingly helpful in building your understanding.

OVERCOMING TRAUMATIC BRAIN INJURY AND LEARNING TO LEARN WITH LIMITED TIME—PAUL KRUCHKO'S STORY

Paul Kruchko with his wife and daughter, who have helped provide the motivation for him to reshape his life.

"I grew up poor, with a lot of domestic turmoil. I barely graduated from high school. Afterward, I enlisted in the army, where I was deployed as an infantryman to Iraq. My vehicle was hit eight out of the twelve times our platoon was ambushed with roadside bombs.

"During my tour, through lucky chance, I met my wonderful wife. Meeting her convinced me to leave the service and start a family. The problem was I didn't know what to do. Worse yet, after returning home I started to have problems with concentration, lack of cognition, and irritability that I had never experienced before. Sometimes I could barely finish a sentence. It was only later that I read about soldiers returning home from Iraq and Afghanistan having issues with undiagnosed traumatic brain injury (TBI).

"I enrolled in a computer and electronics engineering technology program. My TBI was severe enough that at first I even struggled to comprehend fractions.

"But it was a blessing in disguise: The learning was doing something

to my brain. It was as if the mental concentration—difficult though it was—was retooling my mind and helping my brain heal. To me, this process seemed analogous to how I would apply physical effort in the gym, and blood would be forced into my muscles to build strength. In time my mind healed—I graduated with high honors and got a job as a civilian electronics technician.

"I decided to go back to school again for an engineering degree. Mathematics—especially calculus—is far more important when studying engineering than it is when training as a hands-on technician. At this point, my lack of foundation of mathematics from the earliest years of grade school started to catch up with me.

"By this time I was married, a new father, and working full-time. Now the challenge I faced was no longer just basic cognition, but time management. I had only a few hours each day to learn advanced concepts at a far deeper level than I ever had to before. It was only after a few hard knocks (I earned a D in my differential equations class—ouch!) that I started to approach learning in a more strategic way.

"Each semester, I get a copy of the syllabi from my professors and begin reading the textbooks at least two to three weeks before the courses begin. I try to stay at least a chapter ahead of the class, although by the middle of the semester this is often impossible. Practice in problem solving—building chunks—is key. Over my learning career, I gradually developed the following rules, which have allowed me to satisfactorily complete my courses. My objective is a good career that will support my family—these techniques are helping me get there."

Paul's Techniques for Limited Study Time

1. **Read (but don't yet solve) assigned homework and practice exams/quizzes.** *With this initial step I prime my mental pump for learning new concepts—new chunks.*

2. **Review lecture notes** *(attend every lecture as much as possible). One hour of lecture is worth two hours reading the book. I learn far more efficiently if I am faithful in attending lectures and taking detailed notes—not just staring at my watch and waiting for it to be over. I review my notes the following day while the subjects are still fresh in my mind. I've also found that thirty minutes with a professor asking questions is easily worth three hours reading the book.*

3. **Rework example problems presented in lecture notes.** *It never helped me to practice problems given by either the instructor or the textbook that didn't have solutions to provide feedback. With the example problems I already had a step-by-step solution available if necessary. Reworking helps solidify chunks. I use different-colored pens when I study: blue, green, red—not just black. I found that it helps me focus on reading my notes better; things pop out more, instead of blending together into a confusing collage of inexplicable mathematical chaos on the page.*

4. **Work assigned homework and practice exam/quiz questions.** *This builds "muscle memory" chunks for the mind in solving certain types of problems.*

....

{ 5 }

preventing
procrastination:

Enlisting Your Habits ("Zombies") as Helpers

For centuries, arsenic was a popular choice for killers. A sprinkle on your morning toast would cause your painful death within a day. So you can imagine the shock at the forty-eighth meeting of the German Association of Arts and Sciences in 1875, when two men sat in front of the audience and blithely downed more than double a deadly dose of arsenic. The next day the men were back at the conference, smiling and healthy. Analysis of the men's urine showed it was no trick. The men had indeed ingested the poison.[1]

How is it possible to take something so bad for you and stay alive—and even look healthy?

The answer has an uncanny relationship to procrastination. Understanding something of the cognitive psychology of procrastination, just like understanding the chemistry of poison, can help us develop healthy preventatives.

In this and the next chapter, I'm going to teach you the lazy person's approach to tackling procrastination. This means you'll be learning about your inner zombies—the routine, habitual responses your brain falls into as a result of specific cues. These zombie responses are often focused on making the *here and now* better. As you'll see, you can trick some of these zombies into helping you to fend off procrastination when you need to (not all procrastination is bad).[2] Then we'll interleave a chapter where you'll deepen your chunking skills, before we return with a final chapter of wrap-up coverage on procrastination that provides tips, tricks, and handy technological tools.

First things first. Unlike procrastination, which is easy to fall into, willpower is hard to come by because it uses a lot of neural resources. This means that the *last* thing you want to do in tackling procrastination is to go around spraying willpower on it like it's cheap air freshener. You shouldn't waste willpower on procrastination except when absolutely necessary! Best of all, as you will see, you don't need to.

Poison. Zombies. Could it get any better?

Ah yes—there's experimentation! *Bwah hah hah*—what could be more fun?

DISTRACTION AND PROCRASTINATION

"Procrastination is one of our generation's biggest problems. We have so many distractions. I am always thinking, 'Before I start my homework, let me just check my Facebook, Twitter, Tumblr, and e-mail.' Before I even realize it, I have wasted at least an hour. Even after I finally start my homework, I have those distracting websites open in the background.

"I need to find a way just to focus on my studying and homework. I think it depends a lot on my environment and the time. I should not be waiting until the last minute to do everything."

—*A calculus student*

Procrastination and Discomfort

Imagine how your calf muscles would scream if you prepared for a big race by waiting till midnight the night before your first marathon to do your first practice run. In just the same way, *you can't compete in math and science if you just cram at the last minute.*

For most people, learning math and science depends on two things: brief study sessions where the neural "bricks" are laid, and time in between for the mental mortar to dry. This means that procrastination, a terribly common problem for *many* students,[3] is particularly important for math and science students to master.

We procrastinate about things that make us feel uncomfortable.[4] Medical imaging studies have shown that mathphobes, for example, appear to avoid math because even just thinking about it seems to hurt. The pain centers of their brains light up when they contemplate working on math.[5]

But there's something important to note. It was the *anticipation* that was painful. When the mathphobes actually *did* math, the pain disappeared. Procrastination expert Rita Emmett explains: "The dread of doing a task uses up more time and energy than doing the task itself."[6]

Avoiding something painful seems sensible. But sadly, the long-term effects of habitual avoidance can be nasty. You put off studying math, and it becomes even *more* painful to think about studying it. You delay practicing for the SAT or ACT, and on the critical exam day, you choke because you haven't laid the firm neural foundations you need to feel comfortable with the material. Your opportunity for scholarships evaporates.

Perhaps you'd love a career in math and science, but you give up and settle on a different path. You tell others you couldn't hack the math, when the reality was that you had simply let procrastination get the best of you.

Procrastination is a single, monumentally important "keystone" bad habit.[7] A habit, in other words, that influences many important areas of your life. Change it, and a myriad of other positive changes will gradually begin to unfold.

And there's something more—something crucially important. It's easy to feel distaste for something you're not good at. But **the better you get at something, the more you'll find you enjoy it.**

How the Brain Procrastinates

Beep beep beep . . . It's ten A.M. on Saturday, and your alarm clock pulls you from luscious sleep. An hour later you're finally up, coffee in hand, poised over your books and your laptop. You've been meaning to put in a solid day of studying so you can wrap up that math homework that's due on Monday. You also plan to get a start on the history essay, and to look at that confusing chemistry section.

You look at your math textbook. There's a subtle, barely detectable *ouch*. Your brain's pain centers light up as you anticipate looking at the confusing graphs and tangle of strange verbiage. You *really* don't want to be doing math homework now. The thought of spending the next several hours studying math, as you'd planned, makes the idea of opening the book even less pleasant.

You shift your focus from your textbook to your laptop. Hmm, that's more like it. No painful feelings there, just a little dollop of pleasure as you flip open the screen and check your messages. Look at that funny picture Jesse sent . . .

Two hours later, you haven't even started your math homework.

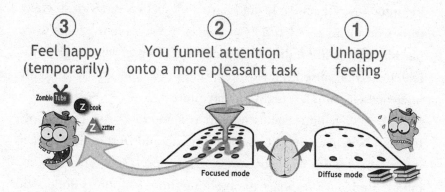

This is a typical procrastination pattern. You think about something you don't particularly like, and the pain centers of your brain light up. So you shift and narrow your focus of attention to something more enjoyable.[8] This causes you to feel better, at least temporarily.

Procrastination is like addiction. It offers temporary excitement and relief from boring reality. It's easy to delude yourself that the most profitable use of any given moment is surfing the web for information instead of reading the textbook or doing the assigned problems. You start to tell yourself stories. For example, that organic chemistry requires spatial reasoning—your weakness—so *of course* you're doing poorly at it. You devise irrational excuses that sound superficially reasonable: *If I study too far ahead of a test, I'll forget the material.* (You conveniently forget the tests in *other* courses you'll be taking during exam time, making it impossible to learn all the material at once.) Only when the semester is ending and you start your desperate cramming for the final exam do you realize that the real reason you are doing so badly in organic chemistry is that you have been continually procrastinating.

Researchers have found that procrastination can even become a source of pride as well as an excuse for doing poorly. "I crammed for the quiz last night after finishing the lab report and the marketing interview. Of course I could have done better. But with so many things on my plate, what do you expect?"[9] Even when people work hard at their studies, they sometimes like to falsely claim they procrastinated because it makes them seem cool and smart: "I finally made myself cram last night for the midterm."

Like any habit, procrastination is something you can simply fall into. You get your procrastination cue and unthinkingly relax into your comfortable procrastination response. Over time, your habitual, zombielike response in obtaining those temporary dollops of pleasure can gradually lower your self-confidence, leaving you with even less of a desire to learn how to work effectively. Procrastinators report higher stress, worse health, and lower grades.[10] As time goes on, the habit can become entrenched. At that point, fixing it can feel hopeless.[11]

CHANGE IS POSSIBLE

"I used to be a procrastinator but I've changed. I had an AP class in high school that really helped get me into gear. My teacher assigned four to six hours of American history homework a night. What I learned is to take it one task at a time. I've found that if I feel like I've accomplished something, it's easier to keep moving forward and stay on track."

—*Paula Meerschaert, freshman, creative writing*

Occasionally you can pull an all-nighter in your studies and still get a decent grade. You can even feel a sort of high when you've finished. Much as with gambling, this minor win can serve as a reward that prompts you to take a chance and procrastinate again. You may even start telling yourself that procrastination is an innate characteristic—a trait that is as much a part of you as your height or the color of your hair. After all, if procrastination were easily fixable, wouldn't you have fixed it by now?

The higher you go in math and science, however, the more important it is to take control of procrastination. Habits that worked in earlier years can turn around and bite you. What I'll show you in these next few chapters is how you can become the master of your habits. *You* should be making your decisions, not your well-meaning, but unthinking, zombies—your habits. As you will see, the strategies for dealing with procrastination are simple. It's just that they aren't intuitively obvious.

Let's return to the story that began this chapter. The arsenic eaters started with tiny doses of arsenic. In tiny doses, arsenic doesn't seem harmful. You can even build up an immunity to its effects. This can allow you to take large doses and look healthy even as the poison is slowly increasing your risk of cancer and ravaging your organs.

In a similar way, procrastinators put off just that one *little* thing. They do it again and again, gradually growing used to it. They can even look healthy. But the long-term effects?

Not so good.

A LITTLE GOES A LONG WAY

"When a student complains of failing and tells me he studied for *ten whole hours* the day before the exam, I answer, 'That's why you failed.' When the student looks at me in disbelief, I say, 'You should have been studying a *little bit* all along.'"

—*Richard Nadel, Senior Instructor of Mathematics,*
Florida International University, Miami, Florida

SUMMING IT UP

- We procrastinate about things that make us feel uncomfortable. But what makes us feel good temporarily isn't
  necessarily good for us in the long run.
- Procrastination can be like taking tiny amounts of poison.
  It may not seem harmful at the time. But the long-term
  effects can be very damaging.

PAUSE AND RECALL

In chapter 4, we learned that it can help to recall material when you are in a physically different location from
where you originally learned it. This helps you become
independent of location cues. Later, you will find yourself thinking more comfortably about the material no
matter where you are—this is often important when
you are being tested.

Let's try this idea now. What were the main ideas of
this chapter? You can recall them where you are currently sitting, but then try recalling the ideas again in a
different room, or better yet, when you are outside.

ENHANCE YOUR LEARNING

1. Have habits of procrastination had an impact on your life? If so, how?

2. What types of stories have you heard other people tell about why they procrastinate? Can you see the holes in some of these stories? What holes are in your own stories about procrastination?

3. List some specific actions you could take that would help you curb habits of procrastination without relying very much on willpower.

ACTIVELY SEEK GOOD ADVICE! INSIGHTS
FROM NORMAN FORTENBERRY, A NATIONAL
LEADER IN ENGINEERING EDUCATION

"When I was a first-year college student, I already knew I wanted to be an engineer, so I signed up for Calculus with Applications instead of the regular calculus being taken by most of my classmates. This was a mistake. Many of the students in this class had already taken calculus in high school and were expanding their knowledge base. So I was at a competitive disadvantage.

"More critical, since far fewer students were in the version of calculus that I was taking, there were few potential study partners. Unlike in high school, there is no premium (indeed there is a penalty) for going it alone in college. Professors in engineering, a field where teamwork is an important professional trait, often assume that you're working with others and design homework accordingly. I squeaked through with a B but always felt that I had an inadequate conceptual and intuitive understanding of the fundamentals of calculus and of the subsequent courses that depended on it. I did lots of studying on my own in a just-in-time fashion for the calculus portions of subsequent classes. But that cost me a lot of time that could have been devoted to other pursuits.

"I am lucky that I made it through to graduation with a bachelor's degree in mechanical engineering, and with the encouragement and mentorship of some peers and my faculty advisor, continued on to graduate school and my doctorate in mechanical engineering. But a point to take to heart from all of this is to ask your peers and teachers for good advice as you choose your classes. Their collective wisdom will serve you well."

{ 6 }

zombies everywhere:

*Digging Deeper to Understand
the Habit of Procrastination*

I n the insightful book *The Power of Habit*, author Charles Duhigg describes a lost soul—Lisa Allen, a middle-aged woman who had always struggled with her weight, who had begun drinking and smoking when she was sixteen, and whose husband had left her for another woman. Lisa had never held a job for more than a year and had fallen deeply into debt.

But in a four-year span, Lisa turned her life around completely. She lost sixty pounds, was working toward a master's degree, stopped drinking and smoking, and was so fit that she ran a marathon.

To understand how Lisa made these changes, we need to understand *habit*.

Habits can be good and bad. Habit, after all, is simply when our brain launches into a preprogrammed "zombie" mode. You will probably not be surprised to learn that chunking, that automatically connected neural pattern that arises from frequent practice, is intimately related to habit.[1] **Habit is an energy saver for us. It al-**

lows us to free our mind for other types of activities. An example of this is backing your car out of the driveway. The first time you do this, you are on hyper-alert. The deluge of information coming at you made the task seem almost impossibly difficult. But you quickly learned how to chunk this information so that before you knew it, all you have to do was think *Let's go,* and you were backing out of the driveway. Your brain goes into a sort of zombie mode, where it isn't consciously aware of everything it is doing.

You go into this habitual zombie mode far more often than you might think. That's the point of the habit—you don't think in a focused manner about what you are doing while you are performing the habit. It saves energy.

Habitual actions can vary in length. They can be brief: seconds-long intervals where you smile absently at a passerby or glance at your fingernails to see whether they are clean. Habits can also take some time—for example, when you go for a run or watch television for a few hours after you get home from work.

Habits have four parts:

1. **The Cue:** This is the trigger that launches you into "zombie mode." The cue may be something as simple as seeing the first item on your to-do list (time to start next week's home-work!) or seeing a text message from a friend (time to daw-dle!). A cue by itself is neither helpful nor harmful. It's the routine—what we do in reaction to that cue—that matters.

2. **The Routine:** This is your zombie mode—the routine, ha-bitual response your brain is used to falling into when it receives the cue. Zombie responses can be harmless, useful, or, in the worst case, so destructive that they defy common sense.

3. **The Reward:** Habits develop and continue because they reward us—give us a dollop of pleasure. Procrastination is an easy habit to develop because the reward—moving your mind's focus to something more pleasant—happens so quickly. But good habits can also be rewarded. Finding ways to reward good study habits in math and science is vital to escaping procrastination.

4. **The Belief:** Habits have power because of your belief in them. For example, you might feel that you'll never be able to change your habit of putting off your studies until late in the day. To change a habit, you'll need to change your underlying belief.

....

"I often find that when I cannot bring myself to start something, if I go for a quick run or do something active first, when I come back to it, it is much easier to start."

—*Katherine Folk, freshman, industrial and systems engineering*

....

Harnessing Your Habits (Your "Zombies") to *Help* You

In this section, we're going to get into the specifics of harnessing your zombie powers of habit to help you avoid procrastination while *minimizing* your use of willpower. You don't want to do a full-scale change of old habits. You just want to overwrite parts of them and develop a few new ones. The trick to overwriting a habit is to look for the pressure point—your reaction to a cue. **The *only* place you need to apply willpower is to change your reaction to the cue.**

To understand that, it helps to go back through the four components of habit and reanalyze them from the perspective of procrastination.

1. **The Cue:** Recognize what launches you into your zombie, procrastination mode. Cues usually fall into one of the following categories: location, time, how you feel, reactions to other people, or something that just happened.[2] Do you look something up on the web and then find yourself web surfing? Does a text message disturb your reverie, taking you ten minutes to get back into the flow of things even when you try to keep yourself on task? The issue with procrastination is that because it's an automatic habit, you are often unaware that you have begun to procrastinate.

    Students often find that developing new cues, such as starting homework as soon as they get home from school or right after their first break from class, are helpful. As procrastination expert Piers Steel, author of *The Procrastination Equation*, points out, "If you protect your routine, eventually it will protect you."[3]

You can prevent the most damaging cues by shutting off your cell phone or keeping yourself away from the Internet for brief periods of time, as when you are working on homework during a twenty-five-minute study session. Freshman actuarial student Yusra Hasan likes to give her phone and laptop to her sister to "watch over," which is doubly clever because a public commitment to study is made in the very act of removing temptation. Friends and family can be helpful if you enlist them.

2. **The Routine:** Let's say that instead of doing your studies, you often divert your attention to something less pain-

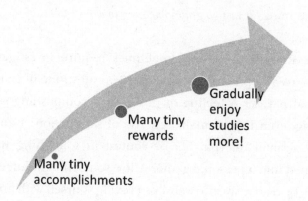

Gradually enjoy studies more!

Many tiny rewards

Many tiny accomplishments

ful. Your brain wants to automatically go into this routine when you've gotten your cue, so this is the pressure point where you must actively focus on rewiring your old habit. **The key to rewiring is to have a plan. Developing a new ritual can be helpful.** Some students make it a habit to leave their smartphone in their car when they head in for class, which removes a potent distraction. Many students discover the value of settling into a quiet spot in the library or, closer to home, the productive effects of simply sitting in a favorite chair at the proper time with all Internet access disconnected. Your plan may not work perfectly at first, but keep at it. Adjust the plan if necessary and savor the victories when your plan works. Don't try to change everything at once. The Pomodoro technique—the twenty-five-minute timer—can be especially helpful in shifting your reaction to cues.

Also, it helps to have something in your stomach when starting particularly difficult tasks. This ensures that you have mental energy for that momentary dollop of willpower

as you are getting started.[4] It also avoids the potential distraction of *I'll just go grab something to eat.* . . .

3. **The Reward:** This can sometimes require investigation.
Why are you procrastinating? Can you substitute in an emotional payoff? A feeling of pride for accomplishing something, even if it is small? A sense of satisfaction? Can you
win a small internal bet or contest in something you've
turned into a personal game? Allow yourself to indulge in a
latte or read a favorite website? Provide yourself with an evening of mindless television or web surfing without guilt?
And will you give yourself a bigger reward for a bigger
achievement—movie tickets, a sweater, or an utterly frivo-
lous purchase?

....

"My boyfriend and I love movies, so as a reward for completing
specific tasks on certain days, he takes me to the movies. This not
only is motivation to study or get homework done, but also has
led me to develop new habits of studying by reinforcing the cue/
routine/reward system."

*—Charlene Brisson, psychology major, accelerated
second-degree nursing program*

....

Remember, habits are powerful because they create neuro-
logical cravings. It helps to add a new reward if you want to
overcome your previous cravings. Only once your brain
starts *expecting* the reward will the important rewiring take
place that will allow you to create new habits.

It's particularly important to realize that giving yourself
even a small "attaboy" or "attagirl" jump-starts the process

of rewiring your brain. This rewiring, sometimes called *learned industriousness*, helps brighten tasks you once thought were boring and uninteresting.[5] As you will find, simply getting into the flow of your work can become its own reward, giving you a sense of productiveness you might not have imagined was possible when you first sat down to begin working. Many people find that setting a reward at a specific time—for example, breaking for lunch with a friend at the deli at noon, or stopping the main tasks at five P.M., gives a solid mini-deadline that can help spur work.

Don't feel bad if you find that you have trouble getting into a "flow" state at first. I sometimes find it takes a few days of drudgery through a few cycles of the Pomodoro technique before flow begins to unfold and I find myself starting to enjoy work in a very new area. Also remember that the better you get at something, the more enjoyable it can become.

4. **The Belief:** The most important part of changing your procrastination habit is the belief that you can do it. You may find that when the going gets stressful, you long to fall back into old, more comfortable habits. **Belief that your new system works is what can get you through.** Part of what can underpin your belief is to develop a new community. Hang out with classmates who have the "can do" philosophy that you want to develop. Developing an encouraging culture with like-minded friends can help us remember the values that, in moments of weakness, we tend to forget.

A powerful approach is **mental contrasting.**[6] In this technique, you think about where you are now and contrast it with what you want to achieve. If you're trying to get into medical school, for example, imagine yourself as a doctor, helping others even as you're preparing for a great vacation that you can actually afford. Once you've got that upbeat image in mind, *contrast it with images of your current life.* Imagine your clunker of a car, your macaroni and cheese dinners, and your mountain of student debts. Yet there's hope!

In mental contrasting, it's the contrast of where you want to be with where you are now, or where you have been, that makes the difference. Placing pictures around your work and living spaces that remind you of where you want to be can help prime your diffuse-mode pump. Just remember to contrast those great images with the real, more mundane life that currently surrounds you, or that you are emerging from. You can change your reality.

ONE BAD DAY CAN SPUR MANY BETTER ONES

"Mental contrasting is great! I've been using this since I was a kid—it's something that people could learn to apply to many different situations.

"I once was stuck for months in Maryland working in a chicken supplier factory in the middle of a hot summer. I made up my mind right there that I was going to school to get my degree. This experience is what I use as my mental contrast. I believe that sometimes all it takes is one bad day to spark an important realization. After that, keeping your focus to find the way out of your current situation is much easier."

—Mike Orrell, junior, electrical engineering

NOW YOU TRY!

Practicing Your Zombie Wrangling

Do you like to check your e-mail or Facebook right when you wake up in the morning? Set a timer for ten minutes of work first thing instead—*then* reward yourself with online time. You will be surprised to see that this tiny exercise in self-control will help empower you over your zombies through the day.

Warning: When you first sit down to try this, some of your zombies will scream as if they want to eat your brain. Tune them out! Part of the point of this exercise is learning to laugh at your zombies' antics as they predictably tell you, *"Just this once* it's okay to check Facebook right now."

Get into the Flow by Focusing on *Process*, Not *Product*

If you find yourself avoiding certain tasks because they make you uncomfortable, there is a great way to reframe things: Learn to focus on *process*, not *product*.

Process means the flow of time and the habits and actions associated with that flow of time—as in, "I'm going to spend twenty minutes working." *Product* is an outcome—for example, a homework assignment that you need to finish.

To prevent procrastination, you want to avoid concentrating on *product*. Instead, your attention should be on building processes—habits—that coincidentally allow you to do the unpleasant tasks that need to be done.

For example, let's say you don't like doing your math homework. So you put off working on the homework. *It's only five problems*, you think. *How hard could that be?*

Deep down, you realize that solving five problems could be a daunting task. It's easier to live in a fantasy world where the five homework problems (or the twenty-page report, or whatever) can be done at the last minute.

Your challenge here is to avoid focusing on the **product**—the solved homework problems. *The product is what triggers the pain that causes you to procrastinate.* Instead, you need to focus on the *process,* the small chunks of time you need over days or weeks, to solve the homework problems or prepare for tests. Who cares whether you finished the homework or grasped key concepts in any one session? The whole point instead is that you calmly put forth your best effort for a short period—the *process.*

The essential idea here is that the zombie, habitual part of your brain *likes* processes, because it can march mindlessly along. It's far easier to enlist a friendly zombie habit to help with a *process* than to help with a *product.*

X MARKS THE SPOT!

"It's a good idea to mark the objective of your daily reading assignment with a bookmark (or Post-it note). This gives immediate feedback on progress—you are more motivated when you can see the finish line!"

—*Forrest Newman, Professor of Astronomy and Physics,*
Sacramento City College

Break Your Work into Bite-Sized Pieces—
Then Work Intently, but Briefly

The "Pomodoro" is a technique that's been developed to help you focus your attention over a short period of time. *Pomodoro* is Italian for "tomato"—Francesco Cirillo, who originally developed this time-management system in the 1980s, used a tomato-shaped timer. In the Pomodoro technique, you set a timer for twenty-five minutes. (You were introduced earlier to this idea in one of the "Now You Try!" challenges in chapter 2.) Once the timer starts ticking, you're on the clock. No sneaking off to web surf, chat on the phone, or instant-message your buddies. What's nice about doing a Pomodoro is that if you're working around friends or family, you can tell them about the technique. Then, if they happen to interrupt you, all you need to do is mention that you're "doing a Pomodoro" or "on the clock," and it gives a friendly reason for them to leave you alone.

You may object that it is *stressful* being under the timer. But researchers have found something fascinating and counterintuitive. If you learn under mild stress, you can handle greater stress much more easily. For example, as researcher Sian Beilock describes in her book *Choke*, golfers who practice putting in front of others aren't fazed later on when they have to perform before an audience in competitions. In the same way, if you get used to figuring things out under a mild time crunch, you are much less likely to choke later, when you are in a high-pressure test-taking situation.[7] Time after time, top performers in fields as different as surgery and computer programming deliberately seek coaches who place them under stress by challenging them and driving them to perform better.[8]

Focusing on *process*, not product, is important in avoiding procrastination. It is the consistent, daily **time** you spend getting into the flow of your studies that matters most. **Focus on doing a Pomodoro—a twenty-five-minute timed work session— not on completing a task.** In a similar way, notice how, in this picture, physicist and surfer dude Garret Lisi is focused on the *moment*—not on the accomplishment of having surfed a wave.

When you first try using the Pomodoro, you will probably be amazed at how often the urge arises to take a quick peek at something non-work-related. But at the same time, you will also be pleased at how easy it is to catch yourself and turn your attention back to your work. Twenty-five minutes is such a brief period that almost any adult or near-adult can focus his attention for that long. And when you are done, you can lean back and savor the feeling of accomplishment.

"One helpful tip is to just get started. This advice sounds rela-tively simple, but once you get off to a good start it is much easier to accomplish something. I like to go to the quiet floor in the library because you can often see other people in the same situation. I learn best by visualizing. If I can see other people working on homework, then I am more inclined to do that myself."

—*Joseph Coyne, junior, history*

The key is, when the distraction arises, which it inevitably will, you want to train yourself to ignore it. One of the single most important pieces of advice I can give you on dealing with procrastination is to *ignore distractions*! Of course, setting yourself up so that distractions are minimal is also a good idea. Many students find that either a quiet space or noise-canceling headphones—or both—are invalu-able when they are really trying to concentrate.

OFF WITH DISTRACTIONS!

"I was born without auditory canals and thus am deaf (I'm a Treacher-Collins mutant). So, when I study, off goes the hearing aid, and I can REALLY focus! I love my handicap! I took an IQ test at the end of first grade. My IQ was 90—well below average. My mom was dismayed. I was elated since I thought I made an A grade. I have no idea what my current IQ is. Now that I can hear, it's probably dropped a few notches. Thank God for on/off switches."

—*Bill Zettler, Professor of Biology, codiscoverer of several viruses, and winner of the Teacher of the Year Award, University of Florida*

How soon should you start again once you've done a Pomodoro? It depends what you're doing. If you're trying to get yourself started on something that's due in many weeks, you may reward yourself with a half hour of guilt-free web surfing. If you're under stress and have a lot due, a two- to five-minute breather may have to do. You may want to alternate your Pomodoro sessions with working sessions that don't make use of a timer. If you find yourself lagging and not working with focus, you can put yourself back on the timer.

In Pomodoro-type timer systems, the *process*, which involves simple focused effort, moves to the forefront. You disconnect from being stuck on any one item and can get into a state of automaticity without concerns about *having* to finish anything.[9] This automaticity appears to allow you to more easily access diffuse-mode capabilities. **By focusing on *process* rather than *product*, you allow yourself to back away from judging yourself (*Am I getting closer to finishing?*) and allow yourself to relax into the flow of the work.** This helps prevent the procrastination that can occur not only when you are studying math and science, but when you are doing the writing that is so important for many different college classes.

Multitasking is like constantly pulling up a plant. This kind of constant shifting of your attention means that new ideas and concepts have no chance to take root and flourish. When you multitask while doing schoolwork, you get tired more quickly. Each tiny shift back and forth of attention siphons off energy. Although each attention switch itself seems tiny, the cumulative result is that you accomplish far less for your effort. You also don't remember as well, you make more mistakes, and you are less able to transfer what little you do learn into other contexts. A typical negative example of multitasking is that on average, students who allow themselves to multitask while studying or sitting in class have been found to receive consistently lower grades.[10]

Procrastination often involves becoming sidetracked on less essential little tasks, such as pencil sharpening, in part because you can still feel the thrill of accomplishment. Your mind is tricking you. That is why keeping an experimental notebook is so important; we'll talk about that soon.

NOW YOU TRY!

Ignorance Is Bliss

Next time you feel the urge to check your messages, pause and examine the feeling. Acknowledge it. Then ignore it.

Practice *ignoring* distractions. It is a far more powerful technique than trying to will yourself to not feel those distractions in the first place.

SUMMING IT UP

- A little bit of work on something that feels painful can ultimately be very beneficial.
- Habits such as procrastination have four parts:
  - The cue
  - The routine
  - The reward
  - The belief
- Change a habit by responding differently to a cue, or even avoiding that cue altogether. Reward and belief make the change long-lasting.
- Focus on the process (the way you spend your time) instead of the product (what you want to accomplish).

- Use the twenty-five-minute Pomodoro to stay productive for brief periods. Then reward yourself after each successful period of focused attention.
- Be sure to schedule free time to nurture your diffuse mode.
- Mental contrasting is a powerful motivating technique—think about the worst aspects of your present or past experiences and contrast these with the upbeat vision of your future.
- Multitasking means that you are not able to make full, rich connections in your thinking, because the part of your brain that helps make connections is constantly being pulled away before neural connections can be firmed up.

PAUSE AND RECALL

If you feel muzzy or featherbrained as you're trying to look away and recall a key idea, or you find yourself rereading the same paragraphs over and over again, try doing a few situps, pushups, or jumping jacks. A little physical exertion can have a surprisingly positive effect on your ability to understand and recall. Try doing something active now, before recalling the ideas of this chapter.

ENHANCE YOUR LEARNING

1. Why do you think the zombie-like, habitual part of your brain might prefer *process* to *product*? What can you do to encourage a process orientation even two years from now, long after you've finished this book?

2. What kind of subtle change could you make in one of your current habits that could help you avoid procrastination?

3. What kind of simple and easy *new* habit could you form that would help you avoid procrastination?

4. What is one of your most troublesome cues that spins you off into a procrastination response? What could you do to react differently to that cue, or to avoid receiving the cue?

MATH PROFESSOR ORALDO "BUDDY" SAUCEDO
ON HOW FAILURE CAN FUEL SUCCESS

Oraldo "Buddy" Saucedo is a highly recommended math professor on RateMy-Professors.com; he is a full-time math instructor for the Dallas County Community College District in Texas. One of his teaching mottos is "I offer opportunities for success." Here, Buddy provides insight into a failure that fueled his success.

"Every once in a while, a student asks me if I have always been smart—this makes me laugh. I then proceed to tell them about my initial GPA at Texas A&M University.

"While writing '4.0' on the whiteboard, I say that I was close to having a 4.0 my first semester. 'Sounds great, right?' I ask, pausing for their reaction. Then I take my eraser and move the decimal point over to the left. It ends up looking like this: '0.4.'

"Yes. It's true. I failed miserably and was kicked out of the university. Shocking, right? But I did return and eventually received both my bachelor's and master's.

"There are a lot of failure-to-success types out there with similar stories. If you've failed in the past, you may not realize how important that it can be in fueling your success.

"Here are some of the important lessons I've learned in my climb to success:

- *You are not your grade; you are better than that. Grades are indicators of time management and a rate of success.*
- *Bad grades do not mean you are a bad person.*
- *Procrastination is the death of success.*
- *Focusing on taking small, manageable steps forward and time management are key.*
- *Preparation is key to success.*

- *We all have a failure rate. You will fail. So control your failures. That is why we do homework—to exhaust our failure rate.*
- *The biggest lie ever is that practice makes perfect. Not true—practice makes you better.*
- *Practice is where you are supposed to fail.*
- *Practice at home, in class, anytime and anywhere—except on the TEST!*
- *Cramming and passing are not success.*
- *Cramming for tests is the short game with less satisfaction and only temporary results.*
- *Learning is the long game with life's biggest rewards.*
- *We should ALWAYS be perpetual learners. Always in ALL WAYS.*
- *Embrace failure. Celebrate each failure.*
- *Thomas Edison renamed his failures: "1,000 ways to NOT create a lightbulb." Rename your own failures.*
- *Even zombies get up and try again!*

"They say experience is the best teacher. Instead, it should be that failure is the best teacher. I've found that the best learners are the ones who cope best with failure and use it as a learning tool."

{ 7 }

chunking versus choking:

*How to Increase Your
Expertise and Reduce Anxiety*

New inventions almost never initially appear in their fully formed glory. Rather, they go through many iterations and are constantly being improved. The first "mobile" phones were about as portable as bowling balls. The first clumsy refrigerators were cranky devices used by breweries. The earliest engines were overbuilt monstrosities with about as much power as today's go-karts.

Enhancements come only after an invention has been out for a while and people have had a chance to mess with it. If you have a working engine on hand, for example, it's a lot easier to improve any particular feature or add new ones. That's how clever innovations such as engine turbocharging arose. Engineers realized they could get more power and bang for the buck by stuffing more air and fuel into the combustion chamber. German, Swiss, French, and American engineers, among many others, raced to tweak and improve the basic idea.

....

Did you remember to skim ahead and check the questions at the end of the chapter to help to help you start building chunks of understanding?

....

How to Build a Powerful Chunk

In this chapter, much as with enhancing and refining inventions, we're going to learn to enhance and refine our chunking skills. Creating a little library of these chunks will help you perform better on tests and solve problems more creatively. These processes will lay the groundwork for you to become an expert at whatever you're working on.[1] (In case you are wondering, our jump in this chapter from procrastination back to chunking is an example of *interleaving*— varying your learning by hopping back after a break to strengthen an approach you've learned earlier.)

Here's a key idea: *Learning fundamental concepts of math and science can be a lot easier than learning subjects that require a lot of rote memorization.* This is not to trivialize the difficulty or importance of memorization. Ask any medical school student preparing for board exams!

One reason that statement is true is that **once you start working on a math or science problem, you'll notice that** *each step you complete signals the next step to you.* Internalizing problem-solving techniques enhances the neural activity that allows you to more easily hear the whispers of your growing intuition. When you know— really *know*—how to solve a problem just by looking at it, you've created a commanding chunk that sweeps like a song through your mind. A library of these chunks gives you an understanding of fundamental concepts in a way nothing else can.

So with that, here we go:

STEPS TO BUILDING A POWERFUL CHUNK

1. **Work a key problem all the way through on paper.** (You should have the solution to this problem available, either because you've already worked it or because it's a solved problem from your book. But don't look at the solution unless you absolutely have to!) As you work through this problem, there should be no cheating, skipping steps, or saying, "Yeah, I've got it" before you've fully worked it out. *Make sure each step makes sense.*

2. **Do another repetition of the problem, paying attention to the key processes.** If it seems a little odd to work a problem again, keep in mind that you would never learn to play a song on the guitar by playing it only once, or work out by lifting a weight a single time.

3. **Take a break.** You can study other aspects of the subject if you need to, but then go do something different. Work at your part-time job, study a different subject,[2] or go play basketball. You need to give your diffuse mode time to internalize the problem.

4. **Sleep.** Before you go to sleep, work the problem again.[3] If you get stuck, *listen to the problem.* Let your subconscious tell you what to do next.

5. **Do another repetition.** As soon as you can the next day, work the problem again. You should see that you are able to solve the problem more quickly now. Your understanding should be deeper. You may even wonder why you ever had any trouble with it. At this point, you can start lightening up on computing each step. *Keep your focus on the parts of the problem that are the most difficult for you.* This continued focus on the hard stuff is called "deliberate practice." Although it can sometimes be tiring, it is one of the most important aspects of productive studying. An alternative or supplement at this point is to see whether you can do a similar problem with ease.)

6. **Add a new problem.** Pick another key problem and begin working on it in the same way that you did the first problem. The solution to this problem will become the second chunk in your chunked library. Repeat steps one through five on this new problem. And after you become comfortable with that problem, move on to another. You will be surprised how even just a few solid chunks in your library can greatly enhance your mastery of the material and your ability to solve new problems efficiently.

7. **Do "active" repetitions.** Mentally review key problem steps in your mind while doing something active, such as walking to the library or exercising. You can also use spare minutes to review as you are waiting for a bus, sitting in the passenger seat of a car, or twiddling your thumbs until a professor arrives in the classroom. This type of active rehearsal helps strengthen your ability to recall key ideas when you are solving homework problems or taking a test.

....

That's it. Those are the key steps to building a chunked library. What you are doing is building and strengthening an increasingly interconnected web of neurons—enriching and strengthening your chunks.[4] This makes use of what is known as the *generation effect*. **Generating (that is, recalling) the material helps you learn it much more effectively than simply rereading it.**

This is useful information, but I can already hear what you're thinking: "I'm spending hours every week just solving all my assigned problems once. How am I supposed to do it four times for one problem?"

In response, I would ask you: What is your real goal? To turn in homework? Or to perform well on the tests that demonstrate mastery of the material and form the basis for most of your course grade? Remember, just solving a problem with the book open in front of you doesn't guarantee you could solve something like it again on a test, and, more important, it doesn't mean that you truly understand the material.

If you are pressed for time, use this technique on a few key problems as a form of deliberate practice to speed and strengthen your learning and to help you speed your problem-solving skills.

THE LAW OF SERENDIPITY

Remember, Lady Luck favors the one who tries. So don't feel overwhelmed with everything you need to learn about a new subject. Instead, focus on nailing down a few key ideas. You'll be surprised at how much that simple framework can help.

The way in which musicians improve their ability to play an instrument can also be applied to learning math in this sense: A master violinist, for example, doesn't just play a musical piece from beginning to end, over and over again. Instead, she focuses on the hardest parts of the piece—the parts where the fingers fumble and the mind becomes confused.[5] You should be like that in your own deliberate practice, focusing and becoming quicker at the hardest parts of the solution procedures you are trying to learn.[6]

Remember, research has shown that the more effort you put into recalling material, the deeper it embeds itself into your memory.[7] Recall, not simple rereading, is the best form of deliberate practice in study. This strategy is also similar to that used by chess masters. These mental wizards internalize board configurations as chunks associated with the best next moves in their long-term memory. Those mental structures help them select their best option for each move in their current game.[8] The difference between lesser-ranked players and grand masters is that grand masters devote far more time to figuring out what their weaknesses are and working to strengthen those areas.[9] It's not as easy as just sitting around and playing chess for fun. But in the end, the results can be far more gratifying.

Remember, retrieval practice is one of the most powerful forms of learning. It is far more productive than simply rereading material.[10] Building a chunked library of ways to solve problems is effective precisely because it is built on methods of retrieval practice. Do not be fooled by illusions of competence. Remember, just staring at material that's already on the page in front of you can fool you into thinking you know it when you actually don't.

When you first start practicing this way, it may feel awkward—as if you're a thirty-year-old sitting down for your first piano lesson. But as you practice, you'll find it gradually coming together more easily and swiftly. Be patient with yourself—as your ease with the material begins to grow, you'll find yourself enjoying it more and more. Is it work? Sure—and so is learning to play the piano with verve and style. But the payoff is well worth the effort!

"CHUNK-PUTERS" ARE GREAT!

"Between being a full-time engineering student and also working a full-time job as an engineering tech, I have too much academic work to keep it all in the forefront of my mind. So my mental trick is to create big chunks for different areas—thermo class, machine design, programming, et cetera. When I need to recall an individual project, I set my current focus aside and reference the desired chunk, which is like a link on my computer desktop. I can either focus in on a specific area or, in diffuse mode, I can look at the complete desktop and find conceptual links between chunks. When I have a clean and organized mental desktop, I can make connections more easily. It increases my mental agility and also allows me to bore deeper into any one topic more easily."

—Mike Orrell, junior, electrical engineering

Hitting the Wall—When Your Knowledge Suddenly Seems to Collapse

Learning doesn't progress logically so that each day just adds an additional neat packet to your knowledge shelf. Sometimes you hit a wall in constructing your understanding. Things that made sense before can suddenly seem confusing.[11]

This type of "knowledge collapse" seems to occur when your mind is restructuring its understanding—building a more solid foundation. In the case of language learners, they experience occasional periods when the foreign language suddenly seems as comprehensible as Klingon.

Remember—it takes time to assimilate new knowledge. You will go through some periods when you seem to take an exasperating step backward in your understanding. This a natural phenomenon that means your mind is wrestling deeply with the material. You'll find that when you emerge from these periods of temporary frustration, your knowledge base will take a surprising step forward.

Getting Your Act Together—Organizing Your Materials

In preparation for a test, have your problems and solutions **neatly organized** so you can go over them quickly. Some students tape handwritten solutions to problems on the relevant pages of their textbook so everything is readily available. (Use painter's masking tape or sticky notes if you plan to later return a book.) The handwritten solution is important because writing by hand increases the odds that what is written will be retained in memory. Alternatively, keep a binder handy with important problems and solutions from the class and the book, so you can go over them again before tests.

WORDS OF WISDOM ABOUT REMEMBERING FROM ONE OF HISTORY'S GREATEST PSYCHOLOGISTS

....

"A curious peculiarity of our memory is that things are impressed better by active than by passive repetition. I mean that in learning by heart (for example), when we almost know the piece, it pays better to wait and recollect by an effort from within, than to look at the book again. If we recover the words in the former way, we shall probably know them the next time; if in the latter way, we shall very likely need the book once more."

—*William James, writing in 1890*[12]

....

Testing Is a Powerful Learning Experience—Give Yourself Little Mini-Tests Constantly

Here's one of the most important reasons to have well-chunked solution methods readily in mind: They **help prevent choking on tests.** Choking—panicking to the point where you freeze—can happen when your working memory is filled to capacity, yet you still don't have enough room for the additional critical pieces you need to solve a problem. Chunking compresses your knowledge and makes room in your working memory for those pieces so you don't go into mental overload so easily. Also, by making more room in your working memory, you have a better chance of remembering important problem-solving details.[13]

Practicing like this is a form of mini-testing. Research has shown that testing isn't just a means of measuring how much you know. **Testing in itself is a powerful learning experience. It changes and adds to what you know, also making dramatic improvements in your ability to retain the material.**[14] This improvement in knowledge because of test taking is called the *testing effect.* It seems to occur because testing strengthens and stabilizes the related neural patterns in your brain. This is precisely what we saw in chapter 4, in the "Practice Makes Permanent" section, with the picture of the darkening patterns in the brain that occurs with repetition.[15]

Improvement because of the testing effect occurs even when the test performance is bad and no feedback is given. When you are self-testing while you are studying, however, you want to do your best to get feedback and check your answers using solutions manuals, the back of the book, or wherever the solution may lie. Also, as we'll discuss later, interaction with peers as well as instructors helps with the learning process.[16]

One reason why building solid chunks is so helpful is that you get plenty of mini-tests in while you are creating those chunks. Studies have shown that students, and even educators, are often shockingly unaware of the benefits of this kind of mini-testing through retrieval practice.[17]

Students think they are just checking how well they're doing when they do a mini-test of their recall. But this active test of recall is one of the *best* learning methods—better than just sitting passively and rereading! By building your library of chunks, with plenty of active practicing at retrieving material over and over again, and testing your recall, you are using some of the best methods possible for learning deeply and well.

NOW YOU TRY!

Build a Mental Solution Library

A key to building mental flexibility and expertise is to *build your library of chunked solution patterns*. This is your rapid-access data bank—always handy in a pinch. This idea isn't just useful for math and science problems—it applies to many areas in life. That's why, for example, it's always a good strategy to look at where the emergency exits are relative to your seat on an airplane or your room in a hotel.

SUMMING IT UP

- *Chunking* means integrating a concept into one smoothly connected neural thought pattern.

- Chunking helps increase the amount of working memory you have available.

- Building a chunked library of concepts and solutions helps build intuition in problem solving.

- When you are building a chunked library, it's important to keep deliberate focus on some of the toughest concepts and aspects of problem solving.

- Occasionally you can study hard and fate deals a bad hand. But remember the Law of Serendipity: If you prepare well by practicing and building a good mental library, you will find that luck will be increasingly on your side. In other words, you guarantee failure if you don't try, but those who consistently give it a good effort will experience many more successes.

> **PAUSE AND RECALL**
>
> What were the main ideas of this chapter? Almost no one can remember a lot of details, and that's okay. You'll be surprised to see how fast your learning progresses if you begin to encapsulate ideas related to what you are learning into a few key chunks.

ENHANCE YOUR LEARNING

1. What does chunking have to do with working memory?

2. Why do you need to solve a problem yourself as part of the chunking process? Why can't you just look at the solution in the back of the book, understand it, and then move on? What are some additional things you can do to help smooth your chunks right before a test?

3. What is the testing effect?

4. Once you've practiced a problem a few times, pause and see if you can sense the feeling of rightness that occurs when you realize what the next step in the solution process is.

5. What is the Law of Serendipity? Think of an example from your own experiences that typifies this idea.

6. How does choking differ from knowledge collapse?

7. Students fool themselves into thinking that they are learning best by rereading the material instead of by testing themselves through recall. How can you keep yourself from falling into this common trap?

NEEL SUNDARESAN, SENIOR DIRECTOR OF
EBAY RESEARCH LABS, ON INSPIRATION AND
THE PATH TO SUCCESS IN MATH AND SCIENCE

Dr. Neel Sundaresan is the creator of the Inspire! program to help students succeed in science, engineering, math, and technology. Some Inspire! scholars—a group of freshmen from disadvantaged backgrounds—recently filed their first patent, which provided a critical intellectual property asset for mobile commerce for eBay. Dr. Sundaresan's own story provides insight into his path to success.

"I did not go to an elite school when I was growing up. In fact, my school was below average—we didn't have the proper teachers for many subjects. But I focused on finding something good in whatever teachers came my way, whether it was an excellent memory or simply an easy smile. This kind of positive attitude helped me appreciate my teachers and keep an open-minded approach toward my classes.

"This same attitude also helped me later in my career. Today, I always actively seek inspiration from the people I work with and for. Whenever I find my spirit bending low, I discover it is because I have stopped looking for people's positive attributes. This means it is time for me to look within and make changes.

"I know this sounds clichéd, but my main inspiration has always been my mother. She was not allowed to study beyond middle school because she would have had to leave her small town to complete high school. She grew up during an exciting but dangerous time in India's struggle for independence. The doors that shut for my mother have made me determined to open doors for others, to help them realize the enormous opportunities that can be so close to their grasp.

"One of my mother's Golden Rules was that 'writing is the foundation of learning.' From grade school through doctoral studies, I have found

immense power in systematically understanding and writing each step of what I really wanted to learn.

"When I was a graduate student, I used to see other students vigorously highlighting steps in proofs or sentences in a passage of a book. I never understood this. Once you highlight, in some sense, you have destroyed the original without any guarantee that you have placed it inside you, where it can flower.

"My own experiences, then, echo the research findings you are learning about in this book. Highlighting should be avoided because, at least in my experience, it provides only an illusion of competence. Retrieval practice is far more powerful. Try to get the main ideas of each page you are reading cemented in your mind before you turn the page.

"I generally liked to work on my more difficult subjects, like math, in the morning, when I was fresh. I still practice this approach today. I have some of my best mental breakthroughs in the bathroom and shower—it's when I take my mind off the subject that the diffuse mode is able to work its magic."

{ 8 }

tools, tips, and tricks

As noted management specialist David Allen points out, "We trick ourselves into doing what we ought to be doing. . . . To a great degree, the highest-performing people I know are those who have installed the best tricks in their lives . . . The smart part of us sets up things for us to do that the not-so-smart part responds to almost automatically, creating behavior that produces high-performance results."[1]

Allen is referring to tricks like wearing exercise clothes to help him get into the mood for exercising or placing an important report by the front door so he can't miss it. One constant refrain I hear from students is that putting themselves in new surroundings—such as the quiet section of a library, which has few interrupting cues—works wonders with procrastination. Research has confirmed that a special place devoted just to working is particularly helpful.[2]

Another trick involves using meditation to help you learn to ignore distracting thoughts.[3] (Meditation is not just for New Age types—a lot of science has revealed its value.[4]) A short, helpful guide to getting started with meditation is *Buddha in Blue Jeans* by Tai Sheridan. It's free as an electronic book and is suitable for people of any faith. And of course there are many meditation apps—just Google around to see what looks workable for you.

A last important trick is to reframe your focus. One student, for example, is able to get himself up at four thirty each weekday morning, not by thinking about how tired he is when he wakes but about how good breakfast will be.

One of the most extraordinary stories of reframing is that of Roger Bannister, the first person to run a mile in less than four minutes. Bannister was a medical school student who couldn't afford a trainer or a special runner's diet. He didn't even have time to run more than thirty minutes a day, squeezed in around his medical studies. Yet Bannister did not focus on all the reasons why he logically had no chance of reaching his goal. He instead refocused on accomplishing his goal in his own way. On the morning he made world history, he got up, ate his usual breakfast, did his required hospital rounds, and then caught a bus to the track.

It's nice to know that there are *positive* mental tricks you can use to your advantage. They make up for some of the negative tricks you can play that either don't work or make things more difficult for you, like telling yourself that you can polish off your homework just before it's due.

It's normal to sit down with a few negative feelings about beginning your work. It's how you handle those feelings that matters. Researchers have found that the difference between slow and fast starters is that the nonprocrastinating fast starters put their negative thinking aside, saying things to themselves like, "Quit wasting time and just get on with it. Once you get it going, you'll feel better about it."[5]

A POSITIVE APPROACH TO PROCRASTINATION

"I tell my students they can procrastinate as long as they follow three rules:

1. No going onto the computer during their procrastination time. It's just too engrossing.

2. Before procrastinating, identify the easiest homework problem. (No solving is necessary at this point.)

3. Copy the equation or equations that are needed to solve the problem onto a small piece of paper and carry the paper around until they are ready to quit procrastinating and get back to work.

"I have found this approach to be helpful because it allows the problem to linger in diffuse mode—students are working on it even while they are procrastinating."

—Elizabeth Ploughman, Lecturer of Physics,
Camosun College, Victoria, British Columbia

Self-Experimentation: The Key to a Better You

Dr. Seth Roberts is a professor emeritus of psychology at the University of California, Berkeley. While learning to perform experiments as a graduate student, he began to experiment on himself. Roberts's first self-experiment involved his acne. A dermatologist had prescribed tetracycline, so Roberts simply counted the number of pimples he had on his face with varying doses of tetracycline. The result? The tetracycline made no difference on the number of pimples he had!

Roberts had stumbled across a finding that would take medicine another decade to discover—that seemingly powerful tetracycline, which has unsafe side effects, doesn't necessarily work on acne. On the other hand, benzoyl peroxide cream *did* work, contrary to what Roberts had originally thought. As Roberts noted, "From my acne research I learned that self-experimentation can be used by non-experts to (a) see if the experts are right and (b) learn something they don't know. I hadn't realized such things were possible."[6] Over the years, Roberts has used his self-experimentation efforts to study his mood, control his weight, and to see the effects of omega-3 on how well his brain functioned.

Overall, Roberts has found that self-experimentation is extremely helpful in testing ideas as well as in generating and developing new hypotheses. As he notes: "By its nature, self-experimentation involves making sharp changes in your life: you don't do X for several weeks, then you do X for several weeks. This, plus the fact that we monitor ourselves in a hundred ways, makes it easy for self-experimentation to reveal unexpected side effects. . . . Moreover, daily measurements of acne, sleep, or anything else, supply a baseline that makes it even easier to see unexpected changes."[7]

Your own self-experimentation, at least to begin with, should be on procrastination. Keep notes on when you don't complete what you had intended to complete, what the cues are, and your zombie-mode habitual reaction to procrastination cues. By logging your reaction, you can apply the subtle pressure you need to change your response to your procrastination cues and gradually improve your working habits. In his excellent book *The Now Habit*, author Neil Fiore suggests keeping a detailed daily schedule of your activities for a week or two to get a handle on where your problem areas are for procrastination.[8] There are many different ways to monitor your behavior. The most important idea here is that keeping a written history over several weeks appears to be critical in helping you make changes. Also, different people function better in certain environments—some need a busy coffee shop, while others need a quiet library. You need to figure out what's best for you.

ISOLATION VERSUS GROUP WORK—TREATING PROCRASTINATION DIFFERENTLY THAN SIMPLY STRUGGLING TO UNDERSTAND

"A tip I have to address procrastination is to isolate yourself from things you know will distract you, including people. Go to a room all alone, or the library so you do not have anything to distract you."

—*Aukury Cowart, sophomore, electrical engineering*

"If I'm struggling in a subject, I find it helpful to study with other people from the same class. That way I can ask questions and we can work together to figure out what we are confused on. Chances are I might know what he or she is confused about and vice versa."

—*Michael Pariseau, junior, mechanical engineering*

Ultimate Zombie Alliance: The Planner-Journal as Your Personal Lab Notebook

The best way for you to gain control of your habits is simple: Once a week, write a brief weekly list of key tasks. Then, each day, write a list of the tasks that you can reasonably work on or accomplish. Try to write this daily task list the evening before.

Why the day before? Research has shown this helps your subconscious to grapple with the tasks on the list so you figure out how to accomplish them.[9] *Writing the list before you go to sleep enlists your zombies to help you accomplish the items on the list the next day.*

Most people use their phone or an online or paper calendar to keep track of important due dates—you are probably using such a system. From your "due date" calendar, write down a weekly to-do list of twenty or fewer key items. Each night, create the next day's daily to-do list from the items on the weekly to-do list. Keep it to five to ten items. Try not to add to the daily list once you've made it unless it involves some unanticipated but important item (you don't want to start creating endless lists). Try to avoid swapping out items on your list.

A zombie without a list is listless.   A happy zombie has a task list!

If you don't write your tasks down in a list, they lurk at the edge of the four-or-so slots in your working memory, taking up valuable mental real estate.

But once you make a task list, it frees working memory for problem solving. Yay! But remember, you must absolutely trust that you will check your planner-journal. If your subconscious doesn't trust you to do that, tasks will begin swirling back up, blocking your working memory.

One more thing. As writing coach Daphne Gray-Grant recommends to her writing clients: "Eat your frogs first thing in the morning." Do the most important and most disliked jobs first, as soon as you wake up. This is incredibly effective.

The following is a day sample I drew up from my own planner-journal. (You can create your own week sample.) Note that there are only six items—some of them are process oriented. For example, I have a paper due to a journal in several months, so I spend a little focused time on most days working toward completing it. A few items are product oriented, but that is only because they are doable within a limited period of time.

NOV. 30

....

- *PNAS* paper (1 hour)
- Go for a walk
- Book (1 section)
- ISE 150: demo prep
- EGR 260: prepare 1 question for final exam
- Finalize upcoming talk

Focus, fun!
Goal finish time for day: 5:00 p.m.

....

Note my reminders: I want to keep my focus on each item when I am working on it, and I want to have fun. I'm well along my list today. I did catch myself getting sidetracked because I forgot to shut down my e-mail. To get myself back into gear, I set a twenty-two-minute Pomodoro challenge using a timer on my computer desktop. (Why twenty-two minutes? Well, why not? I don't have to do the same thing each time. And notice, too, that by moving to Pomodoro mode, I've switched to a process orientation.) None of the items on my list is too big, because I've got other things going on in my day—meetings to go to, a lecture to give. Sometimes I sprinkle a few tasks that involve physical motion on my list, like pulling weeds or sweeping the kitchen. These aren't generally my favorite kinds of tasks, but somehow, because I'm using them as diffuse-mode breaks, I often look forward to them. Mixing other tasks up with your learning seems to make everything more enjoyable and keeps you from prolonged and unhealthy bouts of sitting.

Over time, as I've gained more experience, I've gotten much better at gauging how long it takes to do any given task. You will find yourself improving quickly as you become more realistic about what you can reasonably do in any given time. Some people like to place a number from one to five beside each task, with one being the highest priority and five being an item that would be okay to delay until the next day. Others like to put a star beside high-priority tasks. Some people like to put a box in front of each item so they can check it off. I personally like to put a big black line through each item when I finish it. Whatever floats your boat. You'll be developing a system that works for *you*.

THE FREEDOM OF A SCHEDULE

"To combat procrastination, I make a schedule of everything I have to do. For example, I tell myself, 'Friday, I need to start my paper and then finish it on Saturday. Also, on Saturday, I need to do my math homework. On Sunday, I need to study for my German test.' It really helps me stay organized and practically stress-free. If I don't follow my schedule, then I have twice the amount of work to do the next day, and that's *really* not something I look forward to."

—*Randall Broadwell, mechanical engineering student with a German minor*

Incidentally, if you've tried starting a planner or journal before and not had it work for you, you might try a related technique that has a more obvious reminder function built in: Keep your task list on a chalkboard or whiteboard by your door. And of course, you can still feel that visceral thrill of pleasure every time you check something off your list!

Notice my goal finish time for the day: 5:00 P.M. Doesn't seem right, does it? But it *is* right, and it is one of the most important components of your daily planner-journal. *Planning your quitting time is as important as planning your working time.* Generally, I aim to quit at 5:00 P.M., although when I'm learning something new, it can sometimes be a pleasure to look at it again after I've taken an evening break, just before I go to sleep. And occasionally there is a major project that I'm wrapping up. The 5:00 P.M. quitting time comes about because I have a family I enjoy hanging with, and I like to have plenty of time for a wide variety of reading in the evening. If this seems like too easy a schedule, keep in mind that I rise early and do this six days a week, obviously not something you need to be doing unless your study and work load is extra heavy.

You might think, *Well, yeah, but you're a professor who is past your youthful study days—of course an early quitting time is fine for you!* However, one of my most admired study experts, Cal Newport, used a 5:00 P.M. quitting time through most of his student career.[10] He ended up getting his Ph.D. from MIT. In other words, this method, implausible though it may seem for some, can work for undergraduate and graduate students in rigorous academic programs. Time after time, those who are committed to maintaining healthy leisure time along with their hard work outperform those who doggedly pursue an endless treadmill.[11]

Once you've finished your daily list, you're done for the day. If you find yourself consistently working beyond your planned quitting time, or not finishing the items you've laid out for yourself, your planner-journal will help you catch it and allow you to start making subtle shifts in your working strategy. You have an important goal each day: to jot a few brief notes into your planner-journal for the next day, and a few checkmarks (hopefully) on your current day's accomplishments.

Of course, your life may not lend itself to a schedule with breaks and leisure time. You may be running on fumes with two jobs and too many classes. But however your life is going, try to squeeze a little break time in.

It's important to transform distant deadlines into daily ones. Attack them bit by bit. Big tasks need to be translated into smaller ones that show up on your daily task list. The only way to walk a journey of a thousand miles is to take one step at a time.

NOW YOU TRY!

Planning for Success

Pick a small portion of a task you have been avoiding. Plan where and when you will tackle that portion of the task. Will you go to the library in the afternoon, leaving your cell phone on airplane mode? Will you go into a different room in your house tomorrow evening, leaving your laptop behind and writing by hand to get a start? Whatever you decide, just planning how you will implement what you need to do makes it far more likely that you will succeed in the task.[12]

You may be so used to procrastination and guilt as motivators that it is hard to bring yourself to believe that another system could work. More than that, it may take you a while to figure out how to properly budget your time because you've never before had the luxury of knowing how much time it takes to do a good job without rushing. Chronic procrastinators, as it turns out, tend to see each act of procrastination as a unique, unusual act, a "just this one time" phenomenon that won't be repeated again. Even though it isn't true, it sounds great—so great that you will believe it again and again, because without your planner-journal, there's nothing to counter your thoughts. As Chico Marx once said, "Who you gonna believe, me or your own eyes?"

AVOIDING PROCRASTINATION—INSIGHTS FROM INDUSTRIAL ENGINEERING STUDENT JONATHON MCCORMICK

....

1. I write down assignments in my planner as being due one day before they are really due. That way, I never rush to finish at the last minute, and I still have one full day to think my assignment through before turning it in.

2. I tell my friends that I'm working on my homework. That way, whenever one of them catches me live on Facebook, they'll hold me accountable to the fact that I'm supposed to be doing homework.

3. I have a framed piece of paper with the starting salary of an industrial engineer on my desk. Whenever I can't focus on my task at hand, I look at that and remind myself that it'll pay off in the long run.

....

A little procrastination here and there is unavoidable. But to be effective in learning math and science, you must master your habits. Your zombies must be under your control. Your planner-journal serves as your eyes to keep track of what works. When you first start using a task list, you will often find that you've been too ambitious—there's no way to accomplish it all. But as you fine-tune, you will quickly learn how to set sensible, doable goals.

You may think, *Yes, but what about a time management system? And how do I know what is most important for me to be working on?* That's what the weekly to-do list is all about. It helps you calmly step back, look at the big picture, and set priorities. Setting out your daily list the evening before can also help prevent you from making last-minute decisions that can cost you in the long run.

Do you need to sometimes make changes in your plans because of unforeseen events? Of course! But remember the Law of Serendipity: Lady Luck favors the one who tries. Planning well is part of trying. Keep your eye on the goal, and try not to get too unsettled by occasional roadblocks.

ENLISTING LISTS AND THE IMPORTANCE OF *STARTING*

"I stay organized during the week by making a list of things that need to be done for each day. The list is usually on a lined sheet of paper that I just fold and stick in my pocket. Every day, a couple of times a day, I'll pull it out and double-check that I've done or am going to do whatever is on the agenda for that day. It's nice to be able to cross stuff off the list, especially when it's super long. I have a drawer just full of these folded-up pieces of paper.

"I find it's easier to start one thing, or even a few things at a time, and know that the next time I go to do them, they are already partly done, so there is less to worry about."

—*Michael Gashaj, sophomore, industrial engineering*

Technology Tips: The Best Apps and Programs for Studying

A simple timer plus pen and paper are often the most straightforward tools to avoid procrastination, but you can also make use of technology. Here's a rundown of some of the best student-oriented tools.

NOW YOU TRY!

Best Apps and Programs to Keep on Task

(free versions available unless otherwise noted)

Timers

- The Pomodoro technique (varied prices and resources): http://pomodorotechnique.com/

Tasks, Planning, and Flash Cards

- 30/30—combines timers with a task list: http://3030.binary hammer.com/
- StudyBlue—combines flash cards and notes with text messages when it's time to study again, along with a direct link to the material: http://www.studyblue.com/
- Evernote—one of my personal favorites; very popular for noting task lists and random pieces of information (replaces the little notebook writers have long carried to keep track of their ideas): http://evernote.com/
- Anki—one of the best pure flash card systems, with an excellent spaced repetition algorithm; many excellent premade decks are available for a variety of disciplines: http://ankisrs.net/
- Quizlet.com—allows you to input your own flash cards; you can work with classmates to divide up the duties (free): http://quizlet.com/
- Google Tasks and Calendar: http://mail.google.com/mail/help/tasks/

Limiting Your Time on Time-Wasting Websites

- Freedom—many people swear by this program, available for MacOS, Windows, and Android ($10): http://macfree dom.com/

- StayFocusd—for Google Chrome: https://chrome.google
.com/webstore/detail/stayfocusd/laankejkbhbdhmipfmgcn
gdelahlfoji?hl=en
- LeechBlock—for Firefox: https://addons.mozilla.org/en-us/
firefox/addon/leechblock/
- MeeTimer—for Firefox; tracks and logs where you spend
your time: https://addons.mozilla.org/en-us/firefox/addon/
meetimer/

Cheering Yourself and Others On

- 43 Things—a goal-setting site: http://www.43things.com/
- StickK—a goal-setting site: http://www.stickk.com/
- Coffitivity—modest background noise similar to a coffee
shop: http://coffitivity.com/

Easiest Block of All

- Disable sound notifications on your computer and smartphone!

SUMMING IT UP

- Mental tricks can be powerful tools. The following are
some of the most effective:
    - Put yourself in a place with few interruptions, such as
    a library, to help with procrastination.
    - Practice ignoring distracting thoughts by simply let-
    ting them drift past.
    - If your attitude is troubled, reframe your focus to
    shift attention from the negative to the positive.
    - Realize it's perfectly normal to sit down with a few
    negative feelings about beginning your work.

- Planning your life for "playtime" is one of the most important things you can do to prevent procrastination, and one of the most important reasons to *avoid* procrastination.
- At the heart of procrastination prevention is a reasonable daily to-do list, with a weekly once-over to ensure you're on track from a big-picture perspective.
- Write your daily task list the evening before.
- Eat your frogs first.

{

PAUSE AND RECALL

Close the book and look away. What were the main ideas of this chapter? Remember to congratulate yourself for having finished reading this section—every accomplishment deserves a mental pat on the back!

ENHANCE YOUR LEARNING

1. If it's normal for students to first sit down with a few negative feelings about beginning their work, what can you do to help yourself get over this hurdle?

2. What is the best way for you to gain control of habits of procrastination?

3. Why would you want to write a task list down the evening before you intend to accomplish the tasks?

4. How might you reframe something you are currently perceiving in a negative way?

5. Explain why having a daily quitting time to work toward is so important.

NOW YOU TRY!

Setting Reasonable Goals

I would like the end of this chapter to be the beginning of your own. For the next two weeks, write your weekly goals down at the beginning of each week. Then, each day, write out five to ten small, reasonable daily goals based on your weekly goals. Cross off each item as you complete it, and mentally savor each completed item that you cross off your list. If you need to, break a given task out into a "mini task list" of three small subtasks to help keep yourself motivated.

Remember, part of your mission is to finish your daily tasks by a reasonable time so that you have some guilt-free leisure time for yourself. You are developing a new set of habits that will make your life much more enjoyable!

You can use paper or a notebook, or you can get a chalkboard or whiteboard to post by your door. Whatever you think will work best, that's what you need to do to get started.

COPING WITH LIFE'S TOUGHEST CHALLENGES USING
MAGICAL MATH MARINATION—MARY CHA'S STORY

"My father abandoned my family when I was three weeks old, and my mother died when I was nine. As a result, I did terribly in middle and high school, and while still a teenager, I left my adopted parents' house with $60 to my name.

"I am currently a 3.9 GPA biochemistry major, and I am working toward my goal of going to medical school. I will apply next year.

"What does this have to do with math? Glad you asked!

"When I joined the army at age twenty-five, it was because my life had become financially unmanageable. Joining the army was the best decision of my life—although that's not to say army life was easy. The most difficult period was in Afghanistan. I was happy with my work, but I had little in common with my coworkers. This often left me feeling alienated and alone, so I studied math in my spare time to keep the ideas fresh in my mind.

"My military experience helped me develop good study habits. Not as in stare intently for hours, but as in only got a few minutes here, gotta figure out what I can! Some issue or other was always arising, which meant that I had to do my work in short bursts.

"That's when I accidentally discovered 'magical math marination'—the equivalent of diffuse-mode processing. I'd be stuck on some problems—really stuck, with no clue about what was going on. Then I'd get called out to respond to some explosion or another. While I was out leading the team, or even just sitting quietly, waiting, the back of my mind was simultaneously musing over math problems. I'd come back to my room later that night and everything would be solved!

"Another trick I've discovered is what I call active review. I'll be straightening my hair or showering, but I'm simultaneously reviewing in my head problems that I have already solved. This allows me to keep problems in the forefront of my mind so I won't forget them.

"My process for studying is as follows:

1. Do all the odd problems in a section (or at least enough of each 'type' to complete your understanding).

2. Let the problems marinate.

3. Make sheets with all the important concepts and one example of each type of problem you'd like to add to your toolbox.

4. Before an exam, be able to list everything on your sheets: the subjects, the types of problems within the sections, and the techniques. You'd be surprised by what just being able to list the sections and subjects will do for you, let alone the types of problems and toolbox tricks. This type of verbal recall allows you to recognize types of problems more quickly and have more confidence before you go into the exam.

"When I was younger, I thought that if I didn't get something immediately, it meant I would never be able to get it, or I wasn't smart. That isn't true at all, of course. Now I understand that it's really important to get started on something early, leaving time for it to digest. This leads to stress-free understanding that makes learning a lot more enjoyable."

{ 9 }

procrastination
zombie wrap-up

We've swept through a number of issues related to procrastination in these last few chapters. But here are a few final thoughts that can shed new insight into procrastination.

The Pluses and Minuses of Working Unrelentingly in "The Zone"

A chance meeting of two Microsoft techies at a Friday-night party in 1988 resulted in an exciting solution to a major software stumbling block that Microsoft had basically given up on. The pair left the party to give the idea a shot, firing up a computer and going through the problematic code line by line. Later that evening, it was clear that they were onto something. That *something*, as Frans Johansson describes in his fascinating book *The Click Moment*, turned the nearly abandoned software project into Windows 3.0, which helped turn

Microsoft into the global technology titan it is today.[1] There are times when inspiration seems to erupt from nowhere.

These kinds of rare creative breakthroughs—relaxed moments of insight followed by mentally strenuous, all-out, late-night labor—are very different from a typical day of studying math and science. It's rather like sports: Every once in a while, you have a day of competition when you need to give everything you have under conditions of extraordinary stress. But you certainly wouldn't train every single day under those kinds of conditions.

On days when you are super productive and keep working away long into the night, you may get a lot done—but in subsequent days, if you look at your planner-journal, you may note that you are *less* productive. People who make a habit of getting their work done in binges are much less productive overall than those who generally do their work in reasonable, limited stints.[2] Staying in the zone too long will send you toward burnout.[3]

An impending deadline can ratchet up stress levels, moving you into a zone where the stress hormones can kick in and assist in thinking. But relying on adrenaline can be a dangerous game, because once stress goes too high, the ability to think clearly can disappear. More important, learning math and science for an upcoming examination is very different from finishing a written report by a given due date. This is because math and science demand the development of new neural scaffolds that are different from the social, pictorial, and language-oriented scaffolds that our brains have evolved to excel at. For many people, math- and science-related scaffolds develop slowly, alternating focused-mode and diffuse-mode thinking as the material is absorbed. Especially when it comes to learning math and science, the bingeing excuse, "I do my best work under deadlines," is simply not true.[4]

Remember the arsenic eaters at the beginning of these chapters on procrastination? Back in the 1800s, when arsenic eating took hold in one tiny Austrian population, people ignored how harmful it was long-term, even if tolerance could be built up. It's a little like not recognizing the dangers of procrastination.

Getting a grip on habits of procrastination means acknowledging that something that feels painful at the moment can ultimately be healthy. Overcoming your urge to procrastinate shares much in common with other minor stressors that are ultimately beneficial.

....

"When I am not working, I must relax—not work on something else!"

—Psychologist B. F. Skinner, reflecting on a crucial realization that became a turning point in his career[5]

....

Wise Waiting

We've learned that seemingly good traits can have bad consequences. *Einstellung* in chess—being blocked from seeing a better move because of previously conceived notions—is a fine example. Your focused attention, normally desirable, keeps your mind preoccupied so that it doesn't see better solutions.

Just as focused attention isn't *always* good, seemingly nasty habits of procrastination aren't always bad. Whenever you make up a to-do list, for example, you could be accused of procrastinating on whatever isn't first on your list. A healthy form of procrastination entails learning to *pause and reflect* before jumping in and accomplishing something. You are learning to wait wisely. There is *always* something to be done. Prioritizing allows you to gain big-picture context for your decision making. Sometimes waiting allows a situation to resolve itself.

Pausing and reflecting are key, not only in stopping procrastination but in math and science problem solving in general. You may be surprised to learn that the difference in the way that math experts (professors and graduate students) and math novices (undergraduate students) solve physics problems is that experts are *slower* to begin solving a problem.[6] Experts took an average of forty-five seconds to figure out how they would categorize a problem according to its underlying physics principles. Undergraduates, on the other hand, simply rushed right in, taking only thirty seconds to determine how they should proceed.

Unsurprisingly, the conclusions drawn by the undergraduates were often wrong because their choices were based on superficial appearances rather than underlying principles. It's as if experts took their time to conclude that broccoli is a vegetable and lemon is a fruit, while novices barged in to say that broccoli is a tiny tree while lemons are clearly eggs. Pausing gives you time to access your library of chunks and allows your brain to make connections between a particular problem and the bigger picture.

Waiting is also important in a broader context. When you have difficulty puzzling out a particular math or science concept, it is important not to let frustration take control and dismiss those concepts as too difficult or abstract. In his aptly titled book *Stalling for Time*, FBI hostage negotiator Gary Noesner notes that we could all learn from the successes and failures of hostage negotiation.[7] At the beginning of such situations, emotions run high. Efforts to speed matters along often lead to disaster. Staving off natural desires to react aggressively to emotional provocations allows time for the molecules of emotion to gradually dissipate. The resulting cooler heads save lives.

Emotions that goad you by saying, "Just do it, it feels right," can be misleading in other ways. In choosing your career, for example, "Follow your passion" may be like deciding to marry your favorite movie star. It sounds great until reality rears its head. The proof is in the outcome: **Over the past decades, students who have blindly followed their passion, without rational analysis of whether their choice of career truly was wise, have been *more* unhappy with their job choices than those who coupled passion with rationality.**[8]

All of this relates to my own life. I originally had no passion, talent, or skill in math. But as a result of rational considerations, I became *willing* to get good at it. I worked *hard* to get good at it. And I knew that working hard wasn't enough—*I also had to avoid fooling myself.*

I did get good at math. That opened the door to science. And I gradually got good at that, too. As I got good, the passion also came.

We develop a passion for what we are good at. The mistake is thinking that if we aren't good at something, we do not have and can never develop a passion for it.

Procrastination FAQs

I'm so overwhelmed by how much I've got to do that I avoid thinking about it, even though it only makes my bad situation worse. What can I do when I feel paralyzed by the enormity of the work I need to do?

Write down three "microtasks" that you can do within a few minutes. Remember how Lady Luck favors those who try—just do your best to focus on something worthwhile.

At this point, close your eyes and tell your mind that you have nothing else to worry about, no other concerns, just your first microtask. (I'm not kidding about the "close your eyes" part—remember, that can help disengage you from your previous thought patterns.[9]) You may want to play a Pomodoro game with yourself. Can you get a start on the first few pages of the chapter in twenty-five minutes?

Accomplishing a lot of difficult tasks is like eating a salami. You go slice by slice—bit by bit. Cheer every accomplishment, even the tiniest ones. You're moving ahead!

How long will it take to change my procrastination habits?

Although you will probably see some results right away, it may take about three months of adjustment to get in place a new set of working habits that you like and are comfortable with. Be patient and use common sense—don't attempt to make drastic changes immediately because they may not be sustainable and that may only discourage you more.

My attention tends to hop all over the place, so it's difficult for me to stay focused on the task at hand. Am I doomed to be a procrastinator?

Of course not! Many of my most creative and successful students have overcome ADHD and related attention difficulties using the types of tools I've outlined in this book. You can, too.

If your attention is easily divided, you especially will benefit from tools that help keep you focused on a specific task for a short period of time. These tools include a planner-journal, a whiteboard by your door, a timer, and scheduling and timing apps and programs on your smartphone or computer. All of these tools can help you turn your zombie procrastination habits into zombie "take charge" habits.

INSIGHTS FROM A STUDENT WITH ATTENTION DEFICIT DISORDER

"As a student with attention deficit disorder, I struggle with procrastination on a daily basis, and structure is the only fool-proof way to prevent procrastination. For me, this means writing EVERYTHING down in my planner or notebook—things like as-signment due dates, work hours, and times to hang out with friends. It also means studying in the same area every day and re-moving ALL distractions—for example, turning my cell phone off.

"I now also do things at the same general time every week—my body likes structure and routine; that's why it was so hard at the beginning to break out of my procrastination habits, but it is also why it has been so easy to keep up with new habits after a month of forcing myself into it."

—*Weston Jeshurun, sophomore, undeclared major*

You've told me to use as little as possible of my willpower in dealing with procrastination. But shouldn't I be using my willpower a lot so that I can strengthen it?

Willpower is a lot like muscle. You have to use your muscles to strengthen and develop them over time. But at any given time, your muscles have only so much energy available. Developing and using willpower is a bit of a balancing act.[10] This is why it's often important to pick only one difficult thing at a time that requires self-discipline if you are trying to make changes.

It's easy to get myself to sit down and start my schoolwork. But as soon as I start, I find myself taking quick peeks at Facebook or my e-mail. Before I know it, it's taken me eight hours to do a three-hour task.

The Pomodoro timer is your all-purpose zombie distracter. No one ever said you have to be perfect about overcoming habits of procrastination. All you need to do is keep working to improve your process.

What do you say to the student who procrastinates but refuses to accept his own role and instead blames everyone and everything except himself? Or the student who flunks every test but thinks she knows the materials better than her scores show?

If you find yourself constantly falling into situations where you think, "It's not my fault," something is wrong. Ultimately, you are the captain of your fate. If you aren't getting the grades you'd like, you need to start making changes to steer yourself toward better shores, rather than blaming others.

A number of students have told me over the years that they "really knew the material." They protest that they flunked because they don't test well. Often, the student's teammates tell me the real story: The student does little to no studying. It's sad to say that misplaced self-confidence in one's abilities can sometimes reach almost delusional levels. I'm convinced this is part of why employers like to hire people who are successful in math and science. Good grades in those disciplines are often based on objective data about a student's ability to grapple with difficult material.

It's worth reemphasizing that world-class experts in a variety of disciplines reveal that their path to expertise wasn't easy. They slogged through some tedious, difficult times to get to their current level of expertise where they can glide by and make it all look easy.[11]

NOW YOU TRY!

Practicing Your Zombie Wrangling

Think of a challenge that you have been putting off. What kind of thoughts would help you actually do it? For example, you might think: "It's not really so difficult; it will get easier once I get started; sometimes it's good to do things that I don't enjoy; the rewards are worth it."[12]

SUMMING IT UP

Procrastination is such an important topic that this summary includes key takeaway points from *all* this book's chapters on overcoming procrastination:

- Keep a planner-journal so you can easily track when you reach your goals and observe what does and doesn't work.
- Commit yourself to certain routines and tasks each day.
- Write your planned tasks out the night before, so your brain has time to dwell on your goals to help ensure success.

- Arrange your work into a series of small challenges. Always make sure you (and your zombies!) get lots of rewards. Take a few minutes to savor the feelings of happiness and triumph.
- Deliberately delay rewards until you have finished a task.
- Watch for procrastination cues.
- Put yourself in new surroundings with few procrastination cues, such as the quiet section of a library.
- Obstacles arise, but don't make a practice of blaming all your problems on external factors. If everything is always somebody else's fault, it's time to start looking in the mirror.
- Gain trust in your new system. You want to work hard during times of focused concentration—and also trust your system enough that when it comes time to relax, you actually relax without feelings of guilt.
- Have backup plans for when you still procrastinate. No one is perfect, after all.
- Eat your frogs first.

Happy experimenting!

PAUSE AND RECALL

Close the book and look away. What were the main ideas of this chapter? When you go to bed this evening, try recalling the main ideas again—just before sleep often seems to be a particularly powerful time for setting ideas mentally in mind.

ENHANCE YOUR LEARNING

1. If you have problems with being easily distracted, what are some good approaches to help you prevent procrastination?

2. How would you decide when procrastination is useful and when it is harmful?

3. Where have you noticed that pausing and reflecting before charging forward has been beneficial in your life?

4. If you sit down to work but find yourself frittering away your time, what are some actions you can take to quickly get yourself back on task?

5. Reflect on your way of reacting to setbacks. Do you take active responsibility for your part in those setbacks? Or do you assume a victim's role? What way of responding is ultimately most helpful? Why?

6. Why would those who followed their passion in choosing their careers, without balancing their decision with rational analysis of their choice, be less likely to be happy in those careers?

{ 10 }

enhancing your memory

Joshua Foer was a normal guy. But sometimes normal people can do very unusual things.

A recent college grad, Foer (pronounced "four"), lived with his parents while trying to make a go of being a journalist. He didn't have a great memory; he regularly forgot important dates like his girlfriend's birthday, couldn't recall where he'd put his car keys, and forgot he had food in the oven. And in his work, no matter how hard he tried to catch himself, he still wrote *its* instead of *it's*.

But Foer was amazed to find that some people seemed very different. They could memorize the order of a shuffled deck of playing cards in only thirty seconds, or casually absorb dozens of phone numbers, names, faces, events, or dates. Give these people any random poem, and in minutes, they could recite it to you from memory.

Foer was jealous. These brilliant masters of memory, he thought, must have some unusual way their brains were wired that helped them easily remember prodigious amounts of data.

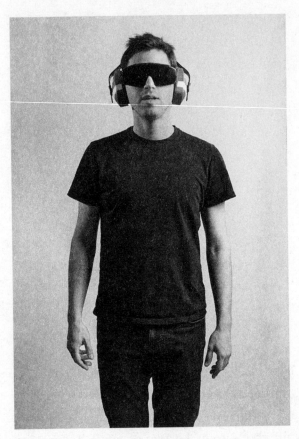

Journalist Josh Foer as he prepares to compete in the U.S. Memory Championships. The ear mufflers and the pinholes in Josh's eye mask help him avoid distraction, which is the competitive memorizer's greatest enemy. This is a firm reminder that it's best to focus without distraction if you really want to put something into memory.

But the memory aces Foer talked to each insisted that their previous, untrained ability to remember was perfectly average. Improbable though it seemed, these people claimed that ancient visualizing techniques were what enabled them to remember so quickly and easily. Anybody can do it, Foer heard repeatedly. *Even you could do it.*[1]

And that goading is how, in one of the most unlikely scenarios Foer could have imagined, he found himself staring at a deck of cards as a top finalist at the U.S. Memory Championships.

"As educators, in our zeal to encourage students to form chunks rather than simply memorize isolated facts, we sometimes give the impression that memorization is unimportant. ('Why should I memorize an equation that I can look up?') But memorization of key facts is essential since it is these facts that form the seeds for the creative process of chunking! The important lesson is that we must continue jiggling and playing mentally with things we have memorized in order to form chunks."

—Forrest Newman, professor of astronomy
and physics, Sacramento City College

Can You Remember Where Your Kitchen Table Is? Your Supersized Visuospatial Memory

It may surprise you to learn that we have *outstanding* visual and spatial memory systems. When you use techniques that rely on those systems, you're not just relying on raw repetition to burn information into your brain. Instead, you're using fun, memorable, creative approaches that make it easier to see, feel, or hear what you want to remember. Even better, these techniques free up your working memory. By grouping things in a sometimes wacky yet logically retrievable fashion, you easily enhance your long-term memory. This can really help take the stress off during tests.

Here's what I mean about your good visual and spatial memory. If you were asked to look around a house you'd never visited before, you would soon have a sense of the general furniture layout, where the rooms were, the color scheme, the pharmaceuticals in the bathroom cupboard (whoa!). In just a few minutes, your mind would acquire and retain thousands of new pieces of information. Even weeks later, you'd still hold far more in your mind than if you'd spent the same amount of time staring at a blank wall. Your mind is built to retain this kind of general information about a place.

The memory tricks used by both ancient and modern memory experts **taps into these naturally supersized visuospatial memorization abilities.** Our ancestors never needed a vast memory for names or numbers. But they *did* need a memory for how to get back home from the three-day deer hunt, or for the location of the plump blueberries on the rocky slopes to the south of camp. These evolutionary needs helped lock in superior "where things are and how they look" memory systems.

The Power of Memorable Visual Images

To begin tapping into your visual memory system, try making a *very* memorable visual image representing one key item you want to remember.[2] For example, here is a picture you could use to remember Newton's second law: $f = ma$. (This is a fundamental relationship relating *force* to *mass* and *acceleration* that only took humans a couple hundred thousand years to figure out.) The letter *f* in the formula could stand for flying, *m* for mule, and *a*, well, that's up to you.

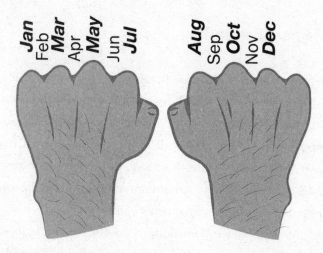

A creative memory device—the months with the projecting knuckles on hands have thirty-one days. As one college calculus student noted: "Oddly enough, with that simple memory tool I doubt I will ever forget which months have thirty-one days—which amazes me. Ten seconds to learn something I've just avoided learning for twenty years because I thought it would be too tedious to sit there and memorize it through repetition."

Part of the reason an image is so important to memory is that images connect directly to your right brain's visuospatial centers.[3] The image helps you encapsulate a seemingly humdrum and hard-to-remember concept by tapping into visual areas with enhanced memory abilities.

The more neural hooks you can build by evoking the senses, the easier it will be for you to recall the concept and what it means. Beyond merely *seeing* the mule, you can *smell* the mule and *feel* the same windy pressure the mule is feeling. You can even *hear* the wind whistling past. The funnier and more evocative the images, the better.

The Memory Palace Technique

The memory palace technique involves calling to mind a familiar place—like the layout of your house—and using it as a sort of visual notepad where you can deposit concept-images that you want to remember. All you have to do is call to mind a place you are familiar with: your home, your route to school, or your favorite restaurant. And voilà! In the blink of an imaginative eye, this becomes the memory palace you'll use as your notepad.

The memory palace technique is useful for remembering unrelated items, such as a grocery list (milk, bread, eggs). To use the technique, you might imagine a gigantic bottle of milk just inside your front door, the bread plopped on the couch, and a cracked egg dribbling off the edge of the coffee table. In other words, you'd imagine yourself walking through a place you know well, coupled with shockingly memorable images of what you might want to remember.

Let's say you are trying to remember the mineral hardness scale, which ranges from 1 to 10 (talc 1, gypsum 2, calcite 3, fluorite 4, apatite 5, orthoclase 6, quartz 7, topaz 8, corundum 9, diamond 10). You can come up with a memory sentence mnemonic: Terrible Giants Can Find Alligators or Quaint Trolls Conveniently Digestible. The problem is that it can still be difficult to remember the sentence. But things become easier if you then add the memory palace. At your front door, there is a terrible giant there, holding a can. Once inside, you find an alligator. . . . You get the idea. If you are studying finance, economics, chemistry, or what-have-you, you'd use the same approach.

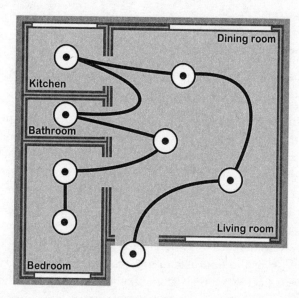

Walk through your memory palace and deposit your memorable images. It's a helpful way to remember lists such as the five elements of a story or the seven steps of the scientific method.

The first time you do this, it will be slow. It takes a bit to conjure up a solid mental image. But the more you do it, the quicker it becomes. One study showed that a person using the memory palace technique could remember more than 95 percent of a forty-to-fifty-item list after only one or two practice mental "walks" where the items were placed on the grounds of the local university.[4] In using the mind this way, memorization can become an outstanding exercise in creativity that simultaneously builds neural hooks for even *more* creativity. What's not to like? (Well, maybe there's one thing not to like: Because this method hooks into your visuospatial system, you do not want to use the memory palace technique when you are doing other spatial tasks, such as driving.[5] The distraction could prove dangerous.)

NOW YOU TRY!

Use the Memory Palace

Top anatomy professor Tracey Magrann applies the memory palace technique to learning the five layers of the epidermis:

"The epidermis has five layers. From deep to superficial, they are the *stratum basale, stratum spinosum, stratum granulosum, stratum lucidum*, and *stratum corneum*. To remember which one is the deepest layer, visualize your basement. That is the *stratum basale*. To get from your basement (deepest layer) to the roof (superficial layer), walk up your basement stairs . . . be careful! They are covered with cactus spines (*stratum spinosum*). That leads you to the kitchen, where someone has spilled granulated sugar all over the floor (*stratum granulosum*). Then you go up-stairs and stop to put on suntan lotion before you go to the roof. The *stratum lucidum* is like a layer of suntan lotion because it protects you from UV rays but is present only on the palms and soles, so that's where you picture yourself applying the lotion. Now you are ready to go to the roof and enjoy a nice corn on the cob (*stratum corneum*)."

Can you think of a way to use the memory palace in your studies?

Songs that help cement ideas in your mind are related to the memory palace technique in that they also make preferential use of the brain's right hemisphere. There are tunes to help you remember the quadratic formula, volume formulas for geometric figures, and many other types of equations. Just Google "quadratic formula" and "song" for examples, or make up your own. Many nursery rhymes use actions along with song to help embed the lyrics (think of "Little Bunny Foo Foo"). Using meaningful motions, from a prance to a jiggle to an itty-bitty hop, can offer even more neural hooks to hold ideas in memory because movement produces sensations that become part of the memory.

These kinds of techniques can be helpful for many things beyond equations, concepts, and grocery lists. Even speeches and presentations—those occasionally petrifying do-or-die experiences—can become much easier when you realize that potentially memorable images can help the key concepts you want to speak about stay in mind. All you need to do is tie the essential ideas you want to talk about to memorable images. See Joshua Foer's masterful TED talk for a demonstration of the memory palace technique for remembering speeches.[6] If you'd like to see how to apply these ideas directly to memorizing formulas, try out the SkillsToolbox .com website for a list of easy-to-remember visuals for mathematical symbols.[7] (For example, the divide symbol "/" is a children's slide.)

Memory aids—whether memorable images, sticky songs, or easily imagined "palaces"—are useful because they help you focus and pay attention when your mind would rather skitter off and do something else. They help remind you that *meaning* is important for remembering, even if the initial meaning is wacky. In short, memorization techniques remind you to make what you learn in your life meaningful, memorable, and fun.

MIND-JOGGING JINGLES

"In my tenth-grade chemistry class, we were introduced to Avogadro's number—6.02214×10^{23}—and none of us could remember it. So one of my friends made up a song about it with a tune borrowed from a Golden Grahams cereal commercial (that turned out to be a much older song called 'Oh, Them Golden Slippers'). So now, thirty years later as an older student, I still remember Avogadro's number because of that song."

—*Malcolm Whitehouse, senior, computer engineering*

TOP TEACHER TRACEY'S MEMORY TIPS

....

"Pacing back and forth, and even having a bit of a snack before-hand, can be helpful when you are memorizing because the brain uses a lot of energy during mental activities. It is also important to make use of multiple areas of the brain when learning. We use the visual cortex of the brain to remember what we see, the auditory cortex for things we hear, the sensory cortex for things we feel, and the motor cortex for things we pick up and move. By using more areas of the brain while learning, we build stronger memory patterns, weaving a tighter web that is less likely to be forgotten during the stress of an exam. For example, in anatomy lab, students should pick up the anatomy models, close their eyes, feel each structure, and say the name of each part out loud. You can skip the senses of smell and taste . . . gotta draw the line somewhere!"

—*Tracey Magrann, Professor of Biological Sciences,*
Saddleback College

....

SUMMING IT UP

- The memory palace technique—placing memorable nudges in a scene that is familiar to you—allows you to dip into the strength of your visual memory system.

- Learning to use your memory in a more disciplined, yet creative manner helps you learn to focus your attention, even as you create wild, diffuse connections that build stronger memories.

- By memorizing material you *understand,* you can internalize the material in a profound way. And you are reinforcing the mental library you need to become a genuine master of the material.

{ **PAUSE AND RECALL**

Close the book and look away. What were the main ideas of this chapter? Tomorrow morning, as you are getting up and beginning your daily "getting out of bed" routine, try to see what you can recall of these key ideas.

ENHANCE YOUR LEARNING

1. Describe an image you could use to help you remember an important equation.

2. Pick any listing of four or more key ideas or concepts from any of your classes. Describe how you would encode those ideas as memorable images and tell where you would deposit them in your memory palace. (For your teacher's sake, you will want to censor some of your more memorable images. As a witty British actress once said, "I don't care what they do, as long as they don't do it in the street and frighten the horses.")

3. Explain the memory palace technique in a way that your grandmother could understand.

SPATIAL ABILITIES CAN BE LEARNED—VISIONARY
ENGINEERING PROFESSOR SHERYL SORBY

Sheryl Sorby is an award-winning engineer whose research interests include designing 3-D computer graphics for visualizing complex behaviors. Here she tells her story.[8]

"Many people erroneously believe that spatial intelligence is a fixed quantity—you either have it or you don't. I am here to say emphatically that this is not the case. In fact, I am living proof that spatial abilities can be learned. I almost left my chosen profession of engineering due to poorly developed spatial skills, but I worked at it, developed the skills, and successfully completed my degree. Because I struggled with spatial skills as a student, I dedicated my career to helping students develop theirs. Virtually all of the students I worked with were able to improve through practice.

"Human intelligence takes many forms, ranging from musical to verbal to mathematical and beyond. An important form is spatial thinking. People with high spatial intelligence can imagine what objects will look like from a different vantage point, or after they have been rotated or sliced in two. In some cases, spatial intelligence might be the ability to figure out the path you would take to get from one place to another, armed only with a map.

"The ability to think in spatial terms has been shown to be important for success in careers such as engineering, architecture, computer science, and many others. Think about the job of air traffic controllers who must imagine the flight paths of several aircraft at a given time, ensuring their paths don't cross. Imagine also the spatial skills required by an auto mechanic to fit parts back into an engine. In recent studies, spatial intelligence has been linked to creativity and innovation. In other words, the

better you are at spatial thinking, the more creative and innovative you will be!

"We have found the reason some students have weak spatial skills is that they likely haven't had many childhood experiences to help develop these skills. Children who spent a good deal of time taking things apart and building them again typically have good spatial ability. Some children who played certain types of sports have good spatial ability. Think of basketball. Players have to imagine the arc necessary for the ball to go into the basket from any place on the court.

"However, even if someone didn't do these kinds of things as a child, it's not too late. Spatial skills can be developed well into adulthood—it just takes practice and patience.

"What can you do? Try accurately sketching an object, and then try sketching it from a different viewpoint. Play 3-D computer games. Put together 3-D puzzles (you may have to start with 2-D puzzles first!). Put away your GPS and try to navigate with a map instead. Above all else, don't give up—instead, just keep working on it!"

{ 11 }

more memory tips

Create a Lively Visual Metaphor or Analogy

One of the best things you can do to not only remember but *understand* concepts in math and science is to **create a metaphor or analogy for it**—often, the more visual, the better.[1] A metaphor is just a way of realizing that one thing is somehow similar to another. Simple ideas like one geography teacher's description of Syria as shaped like a bowl of cereal and Jordan as a Nike Air Jordan sneaker can stick with a student for decades.

If you're trying to understand electrical current, it can help to visualize it as water. Similarly, electrical voltage can "feel like" pressure. Voltage helps push the electrical current to where you want it to go, just like a mechanical pump uses physical pressure to push real water. As you climb to a more sophisticated understanding of electricity, or whatever topic you are concentrating on, you can revise your metaphors, or toss them away and create more meaningful ones.

If you are trying to understand the concept of limits in calculus, you might visualize a runner heading for the finish line. The closer the runner gets, the slower he goes. It's one of those slo-mo camera shots where the runner is never quite able to reach the ribbon, just as we might not quite be able to get to the actual limit. Incidentally, the little book *Calculus Made Easy*, by Silvanus Thompson, has helped generations of students master the subject. Sometimes textbooks can get so focused on all the details that you lose sight of the most important, big-picture concepts. Little books like *Calculus Made Easy* are good to dip into because they help us focus in a simple way on the most important issues.

It's often helpful to pretend *you* are the concept you are trying to understand. Put yourself in an electron's warm and fuzzy slippers as it burrows through a slab of copper, or sneak inside the x of an algebraic equation and feel what it's like to poke your head out of the rabbit hole (just don't let it get exploded with an inadvertent "divide by zero").

MOONBEAMS AND SCHOOL DREAMS

"I always study before I go to bed. For some reason, I usually dream about the material I just studied. Most times these 'school dreams' are quite strange but helpful. For instance, when I was taking an operations research class, I would dream I was running back and forth between nodes, physically acting out the shortest path algorithm. People think I'm crazy, but I think it's great; it means I don't have to study as much as other people do. I guess these dreams involve my subconsciously making metaphors."

—Anthony Sciuto, senior, industrial and systems engineering

In chemistry, compare a cation with a cat that has paws and is therefore "pawsitive," and an anion with an onion that is negative because it makes you cry.

Metaphors are never perfect. But then, *all* scientific models are just metaphors, which means they also break down at some point.[2] But never mind that—metaphors (and models!) are vitally important in giving a physical understanding of the central idea behind the mathematical or scientific process or concept that you are trying to understand. Interestingly, metaphors and analogies are useful for getting people out of *Einstellung*—being blocked by thinking about a problem in the wrong way. For example, telling a simple story of soldiers attacking a fortress from many directions at once can open creative paths for students to intuit how many low-intensity rays can be effectively used to destroy a cancerous tumor.[3]

Metaphors also help glue an idea in your mind, because they make a connection to neural structures that are already there. It's like being able to trace a pattern with tracing paper—metaphors at least help you get a sense of what's going on. If there's a time when you can't think of a metaphor, just put a pen or pencil in your hand and a sheet of paper in front of you. Whether using words or pictures, you will often be amazed at what just noodling about for a minute or two will bring.

METAPHORS AND VISUALIZATION IN SCIENCE

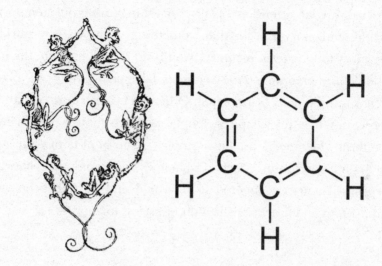

Metaphors and visualization—being able to see something in your mind's eye—have been uniquely powerful in helping the scientific and engineering world move forward.[4] In the 1800s, for example, when chemists began to imagine and visualize the miniature world of molecules, dramatic progress began to be made. Here is a delightful illustration of monkeys in a benzene ring from an insider spoof of German academic chemical life, printed in 1886.[5] Note the single bonds with the monkeys' hands and the double bonds with their tiny tails.

Spaced Repetition to Help Lodge Ideas in Memory

Focusing your attention brings something into your temporary working memory. But for that "something" to move from working memory to long-term memory, two things should happen: the idea should be **memorable** (*there's a gigantic flying mule braying f = ma on my couch!*), and it must be **repeated**. Otherwise, your natural metabolic processes, like tiny vampires, simply suck away faint, newly forming patterns of connections. This vampiric removal of faint patterns is actually a good thing. Much of what goes on around you is basically trivial—if you remembered it all, you'd end up like a hoarder, trapped in an immense collection of useless memories.

If you don't make a point of repeating what you want to remember, your "metabolic vampires" can suck away the neural pattern related to that memory before it can strengthen and solidify.

Repetition is important; even when you make something memorable, repetition helps get that memorable item firmly lodged in long-term memory. But how many times should you repeat? How long should you wait between repetitions?[6] And is there anything you can do to make the repetition process more effective?

Research has given us helpful insight. Let's take a practical example. Say you want to remember information related to the concept of *density*—namely that it is symbolized by a funny-looking symbol, ρ, which is pronounced "row," and that it is measured in standard units of "kilograms per cubic meter."

How can you conveniently and effectively cement this information into memory? (You know now that placing small chunks of information like this in your long-term memory helps gradually build your big-picture understanding of a subject.)

You might take an index card and write "ρ" on one side and the remaining information on the other. **Writing appears to help you to more deeply encode (that is, convert into neural memory structures) what you are trying to learn.** While you are writing out "kilograms per cubic meter," you might imagine a shadowy kilogram (just feel that mass!) lurking in an oversized piece of luggage that happens to be a meter on each side. The more you can turn what you are trying to remember into something memorable, the easier it will be to recall. You will want to say the word and its meaning aloud, to start setting auditory hooks to the material.

Next, just look at the side of the card with the "ρ" on it and see whether you can remember what's on the other side of the card. If you can't, flip it over and remind yourself of what you are supposed to know. If you can remember, put the card away.

Now do something else—perhaps prepare another card and test yourself on it. Once you have several cards together, try running through them all to see if you can remember them. (This helps you interleave your learning.) Don't be surprised if you struggle a bit.

Once you've given your cards a good try, put them away. Wait and take them out again before you go to sleep. Remember that sleep is when your mind repeats patterns and pieces together solutions.

Briefly repeat what you want to remember over several days; perhaps for a few minutes each morning or each evening, change the order of your cards sometimes. Gradually extend the times between repetitions as the material firms itself into your mind. By increasing your spacing as you become more certain of mastery, you will lock the material more firmly into place.[7] (Great flash card systems like Anki have built in algorithms that repeat on a scale ranging from days to months.)

Interestingly, one of the best ways to remember people's names is to simply try to retrieve the people's names from memory at increasing time intervals after first learning the name.[8] Material that you do not review is more easily discounted or forgotten. Your metabolic vampires suck away the links to the memories. This is why **it's wise to be careful about what you decide to skip when reviewing for tests. Your memory for related but nonreviewed material can become impaired.**[9]

SPACED REPETITION— USEFUL FOR BOTH STUDENTS AND PROFESSORS!

....

"I have been advising my students to do spaced repetition over days and weeks, not just in my analytic courses, but also in my History of Ancient Engineering course. When memorizing strange names and terms, it's always best to practice over several days. In fact, that's precisely what I do when I'm preparing for lectures— repeat the terms out loud over a period of several days, so they roll easily off my tongue when I say them in class."

—*Fabian Hadipriono Tan, Professor of Civil Engineering,*
The Ohio State University

....

NOW YOU TRY!

Create a Metaphor to Help You Learn

Think of a concept you are learning now. Is there another process or idea in a completely different field that somehow seems similar to what you are studying? See if you can come up with a helpful metaphor. (Bonus points if there's a touch of silliness!)

Create Meaningful Groups

Another key to memorization is to create meaningful groups that simplify the material. Let's say you wanted to remember four plants that help ward off vampires—garlic, rose, hawthorn, and mustard. The first letters abbreviate to GRHM, so all you need to do is remember the image of a GRAHAM cracker. (Retrieve your cracker from the kitchen table of your memory palace, dust off the vowels, and you're good to go.)

It's much easier to remember numbers by associating them with memorable events. The year 1965 might be when one of your relatives was born, for example. Or you can associate numbers with a numerical system that you're familiar with. For example, 11.0 seconds is a good running time for the 100-meter dash. Or 75 might be the number of knitting stitches cast onto a needle for the ski hats you like to make. Personally, I like to associate numbers with the feelings of when I was or will be at a given age. The number 18 is an easy one—that's when I went out into the world. By age 104, I will be an old but happy great-granny!

Many disciplines use **memorable sentences** to help students memorize concepts; the first letter of each word in the sentence is also the first letter of each word in a list that needs to be memorized.

Medicine, for example, is laden with memorable mnemonics, among the cleaner of which are "Some Lovers Try Positions that They Can't Handle" (to memorize the names of the carpal bones of the hand) and "Old People from Texas Eat Spiders" (for the cranial bones).

Another example is for the increases-by-ten structure of the decimal system: King Henry died while drinking chocolate milk. This translates to kilo—1,000; hecto—100; deca—10; "while" represents 1; deci—0.1; centi—0.01; milli—0.001.

Time after time, these kinds of memory tricks prove helpful. If you're memorizing something commonly used, see whether someone's come up with a particularly memorable memory trick by searching it out online. Otherwise, try coming up with your own.

BEWARE OF MISTAKING A MEMORY TRICK FOR ACTUAL KNOWLEDGE

"In chemistry we have the phrase *skit ti vicer man feconi kuzin*, which has the cadence of a rap song. It represents the first row of the transition metals on the periodic table (Sc, Ti, V, Cr, Mn, Fe, Co, Ni, Cu, Zn). Then, the rest of the transition metals can be placed on a blank periodic table by other memory tricks. For example, students remember to place Ag (silver) and Au (gold) in the same vertical group as Cu (copper) since copper, silver, and gold are all used to make coins.

"Unfortunately, some students come away thinking that's the *reason* these metals are in the same vertical column—because they are used to make coins. The real reason has to do with similarities in chemical properties and valences.

"This is an example of how students sometimes mistake a memory trick for actual knowledge. Always be wary of confusing *what is truly going on* with the *metaphor* you are using to help your memory."

—William Pietro, Professor of Chemistry,
York University, Toronto, Ontario

Create Stories

Notice that the groups mentioned previously often create meaning through story, even though the story might be short. Poor King Henry shouldn't have drunk that chocolate milk! Storytelling in general has long been a vitally important way of understanding and retaining information. Professor Vera Pavri, a historian of science and technology at York University, tells her students not to think of lectures as lectures but as stories where there is a plot, characters, and overall purpose to the discussion. The best lectures in math and science are often framed like thrillers, opening with an intriguing problem that you just *have* to figure out. If your instructor or book doesn't present the material with a question that leaves you wanting to find the answer, see if you can find that question yourself—then set about answering it.[10] And don't forget the value of story as you create memory tricks.

WRITE ON!

"The number one thing I stress when students come to see me is that there is a direct connection between your hand and your brain, and the act of rewriting and organizing your notes is essential to breaking large amounts of information down into smaller digestible chunks. I have many students who prefer to type their notes in a Word document or on slides, and when these students are struggling, the first thing I recommend is to quit typing and start writing. In every case, they perform better on the next section of material."

—Jason Dechant, Ph.D., Course Director,
Health Promotion and Development,
School of Nursing, University of Pittsburgh

Muscle Memory

We've already mentioned that writing out a card by hand appears to help cement ideas in the mind. Although there is little research in this area,[11] many educators have observed that there seems to be a muscle memory related to writing by hand. For example, when you first stare at an equation, it can appear utterly meaningless. But if you thoughtfully write the equation out several times on a sheet of paper, you will be startled by how the equation will begin to take life and meaning in your mind. In a related vein, some learners find that reading problems or formulas aloud helps them understand better. Just be wary of exercises like writing an equation out a hundred times by hand. The first few times may give you value, but after a while, it simply becomes a rote exercise—the time could be better spent elsewhere.

TALK TO YOURSELF

"I often tell my students to talk to themselves instead of just highlighting and rereading. They look at me quizzically, like I am absolutely insane (which could be true). But I have had many students come back to me later and say that it really works and that this is now one of their study tools."

—Dina Miyoshi, Assistant Professor of Psychology,
San Diego Mesa College

Real Muscle Memory

If you *really* want to boost your memory as well as your general ability to learn, it seems one of the best ways to do it is to exercise. Several recent experiments in both animals and humans have found that regular exercise can make a substantive improvement in your memory and learning abilities. Exercise, it seems, helps create new neurons in areas that relate to memory. It also creates new signaling pathways.[12] It seems that different types of exercise—running or walking, for example, versus strength training—may have subtly different molecular effects. But both aerobic and resistance exercise exert similarly powerful results on learning and memory.

Memory Tricks Help You Become an Expert More Quickly

Here's the bottom line. By using mental pictures instead of words to remember things, you can leap more easily into expert status. In other words, *learning to process ideas visually in math and science is a powerful way to become a master of the material*.[13] And using other memory tricks can greatly enhance your ability to learn and retain the material.

Purists might sniff that using oddball memorization gimmicks isn't really learning. But research has shown that students who use these types of tricks outperform those who don't.[14] In addition, imaging research on how people become experts shows that such memory tools speed up the acquisition of both chunks and big-picture templates, helping transform novices to semiexperts much more quickly—even in a matter of weeks.[15] Memory tricks allow people to expand their working memory with easy access to long term memory.

What's more, the memorization process itself becomes an exercise in creativity. The more you memorize using these innovative techniques, the more creative you become. This is because you are building wild, unexpected possibilities for future connections early on, even as you are first internalizing the ideas. The more you practice using this type of "memory muscle," the more easily you will be able to remember. Where at first it may take fifteen minutes to build an evocative image for an equation and embed it in, say, the kitchen sink of your memory palace, it can later take only minutes or seconds to perform a similar task.

You will also realize that as you begin to internalize key aspects of the material, taking a little time to commit the most important points to memory, you come to understand it much more deeply. The formulas will mean far more to you than they would if you simply looked them up in a book. And you'll be able to sling those formulas around more proficiently on tests and in real-world applications.

One study of how actors memorize their scripts showed that they avoid verbatim memorization. Instead, they depend on an understanding of the characters' needs and motivations in order to remember their lines.[16] Similarly, the most important part of your memorization practices is to *understand* what the formulas and solution steps really mean. Understanding also helps a lot with the memorization process.

You may object and say that you're not creative—that an equation or theory could hardly have its own grandiose motivations or persnickety emotional needs to help you understand and remember it. But remember that inner two-year-old. *Your childlike creativity is still there—you just need to reach out to it.*

MEMORY TRICKS *WORK*

....

"On top of working toward my engineering degree, I am in the process of getting my paramedic license (only two months left!) and have to memorize a large selection of drugs and dosages for both adult and pediatric patients. At first, this seemed overwhelming, especially since there will be lives at stake. But I quickly found little tricks that made learning easy. Take, for example, the drug furosemide, also called Lasix, which draws fluid out of the body. The dose I needed to remember was 40 milligrams. This to me was a godsend, as the numbers 4–0 appeared to me in the word (4–0 semide = furosemide). It is things like this that truly can cement ideas and knowledge in our heads. I now don't ever have to even think twice about it. Truly remarkable."

—William Koehler, sophomore, mechanical engineering

NOW YOU TRY!

Songs to Help You Learn

Make up a song to remember an identity, integral, or scientific formula that you need for class. Having some of these important concepts memorized, through whatever trick you need, will make more complicated problems easier and faster to solve.

SUMMING IT UP

- Metaphors can help you learn difficult ideas more quickly.
- Repetition is critical in allowing you to firm up what you want to remember before the ideas fade away.
- Meaningful groups and abbreviations can allow you to simplify and chunk what you are trying to learn so you can store it more easily in memory.
- Stories—even if they are just used as silly memory tricks—can allow you to more easily retain what you are trying to learn.
- Writing and saying what you are trying to learn seems to enhance retention.

- Exercise is powerfully important in helping your neurons to grow and make new connections.

> **PAUSE AND RECALL**
>
> Remember how important it can be to sometimes think of what you are learning in a place different from where you learned it. Try that technique again as you recall the key ideas of this chapter. People sometimes recollect the *feel* of the place where they were studying— even the cushiony feel of the armchair, or the particular music or picture on the wall in the coffee shop where they were sitting—to help cue a memory.

ENHANCE YOUR LEARNING

1. Take a piece of paper and doodle to create a visual or verbal metaphor for a concept you are trying to understand now in math or science.

2. Look at a chapter in a book you are reading in math or science. Create a question about that material that would make you want to learn more about it.

3. Just before you to go sleep, review something mentally that you are trying to learn. To boost this process, review it yet again when you first wake up.

{ 12 }

learning to appreciate your talent

Work toward an Intuitive Understanding

We can learn a lot about how to do math and science from sports. In baseball, for example, you don't learn how to hit in one day. Instead, your body perfects your swing from plenty of repetition over a period of years. Smooth repetition creates muscle memory, so that your body knows what to do from a single thought—one chunk— instead of having to recall all the complex steps involved in hitting a ball.[1]

In the same way, once you understand *why* you do something in math and science, you don't have to keep reexplaining the *how* to yourself every time you do it. It's not necessary to go around with 100 beans in your pocket and to lay out 10 rows of 10 beans again and again so that you *get* that $10 \times 10 = 100$. At some point, you just know it from memory. For example, you memorize the idea that you simply add exponents—those little superscript numbers—when multiplying numbers that have the same base ($10^4 \times 10^5 = 10^9$). If you use the procedure a lot, by doing many different types of problems, you will find that you understand both the *why* and the *how* behind the procedure far better than you do after getting a conventional explanation from a teacher or book. The greater understanding results from the fact that *your* mind constructed the patterns of meaning, rather than simply accepting what someone else has told you. *Remember—people learn by trying to make sense out of information they perceive. They rarely learn anything complex simply by having someone else tell it to them.* (As math teachers say, "Math is not a spectator sport.")

Chess masters, emergency room physicians, fighter pilots, and many other experts often have to make complex decisions rapidly. They shut down their conscious system and instead rely on their well-trained intuition, drawing on their deeply ingrained repertoire of chunks.[2] At some point, self-consciously "understanding" why you do what you do just slows you down and interrupts flow, resulting in worse decisions.

Teachers and professors can inadvertently get too caught up in following rules. In an intriguing study that illustrates this, six people were filmed doing CPR, only one of whom was a professional paramedic.[3] Professional paramedics were then asked to guess who was the real paramedic. Ninety percent of these "real deal" expert paramedics chose correctly, remarking along the lines of "he seemed to know what he was doing."[4] CPR instructors, on the other hand, could pick the real paramedic out of the lineup only 30 percent of the time. These overly picky theoreticians criticized the real experts in the films for issues such as not taking the time to stop and measure where to put their hands. Precise rule following had come to mean more to the instructors than practicality.

Once you understand *why* you do something in math and science, you shouldn't keep reexplaining the *how*. Such overthinking can lead to choking.

No Need for Genius Envy

Just as Olympic athletes don't build their athletic prowess simply by spending a few hours jogging on the weekends or lifting a few weights in their spare time, chess grand masters don't construct their neural structures through last-minute cramming. Instead, their knowledge base is gradually built over time and with plenty of practice that builds their understanding of big-picture context. Practice like this places the memory traces prominently in the warehouse of long-term memory, where the neural pattern can be quickly and easily accessed when needed.[5]

Let's return to chess master Magnus Carlsen—that fast-thinking genius of speed chess as well as regular chess. Carlsen has an extraordinary grasp of the patterns of thousands of previously played chess games—he can look at the arrangement of an endgame on a chess board and instantly tell you which of more than ten thousand games of past centuries it was drawn from. In other words, Carlsen has created a vast chunked library of potential solution patterns. He can quickly riffle through the chunks to see what others have done when faced with situations similar to what he is facing.[6]

Carlsen isn't unusual in what he is doing, although he does it better than all but a very few past and present chess players. It is typical for grand masters to spend at least a decade practicing and studying to learn thousands of memory chunk patterns.[7] These readily available patterns allow them to recognize the key elements in any game setup much more quickly than amateurs; they develop a professional eye so they can rapidly intuit the best course of action in any situation.[8]

But wait. Aren't chess masters and people who can multiply six-digit numbers in their heads simply exceptionally gifted? Not necessarily. I'm going to tell it to you straight—sure, intelligence matters. Being smarter often equates to having a larger working memory. Your hot rod of a memory may be able to hold nine things instead of four, and you latch onto those things like a bulldog, which makes it easier to learn math and science.

But guess what? It also makes it more difficult for you to be creative.

How is that?

It's our old friend and enemy—*Einstellung.* The idea you already are holding in mind blocks you from fresh thoughts. A superb working memory can hold its thoughts so tightly that new thoughts can't easily peek through. Such tightly controlled attention could use an occasional whiff of ADHD-like fresh air—the ability, in other words, to have your attention shift even if you don't want it to shift. Your ability to solve complex problems may make you overthink simple problems, going for the convoluted answer and overlooking the simple, more obvious solution. Research has shown that smart people can have more of a tendency to lose themselves in the weeds of complexity. People with less apparent intellectual horsepower, on the other hand, can cut more easily to simpler solutions.[9]

IT'S NOT *WHAT* YOU KNOW; IT'S *HOW* YOU THINK

"Experience has shown me an almost inverse correlation between high GRE scores and ultimate career success. Indeed, many of the students with the lowest scores became highly successful, whereas a surprising number of the 'geniuses' fell by the wayside for some reason or other."[10]

—Bill Zettler, Ph.D., Professor of Biology, longtime academic advisor, and winner of the Teacher of the Year Award, University of Florida, Gainesville, Florida

---

If you are one of those people who can't hold a lot in mind at once— you lose focus and start daydreaming in lectures, and have to get to someplace quiet to focus so you can use your working memory to its maximum—well, welcome to the clan of the creative. Having a somewhat smaller working memory means you can more easily generalize your learning into new, more creative combinations. Because your working memory, which grows from the focusing abilities of the prefrontal cortex, doesn't lock everything up so tightly, you can more easily get input from other parts of your brain. These other areas, which include the sensory cortex, not only are more in tune with what's going on in the environment, but also are the source of dreams, not to mention creative ideas.[11] You may have to work harder sometimes (or even much of the time) to understand what's going on, but once you've got something chunked, you can take that chunk and turn it outside in and inside round—putting it through creative paces even you didn't think you were capable of!

Here's another point to put into your mental chunker: Chess, that bastion of intellectuals, has some *elite players with roughly average IQs*. These seemingly middling intellects are able to do better than some more intelligent players because they practice more.[12] That's the key idea. Every chess player, whether average or elite, grows talent by practicing. **It is the *practice*—particularly deliberate practice on the toughest aspects of the material—that can help lift average brains into the realm of those with more "natural" gifts.** Just as you can practice lifting weights and get bigger muscles over time, you can also practice certain mental patterns that deepen and enlarge in your mind. Interestingly, it seems that practice may help you expand your working memory. Researchers on recall have found that doing exercises to repeat longer and longer strings of digits backward seems to improve working memory.[13]

Gifted people have their own set of difficulties. Sometimes highly gifted kids are bullied, so they learn to hide or suppress their giftedness. This can be difficult to recover from.[14] Smarter people also sometimes struggle because they can so easily imagine every complexity, good and bad. Extremely smart people are more likely than people of normal intelligence to procrastinate because it always worked when they were growing up, which means they are less likely to learn certain critical life skills early on.

Whether you are naturally gifted or you have to struggle to get a solid grasp the fundamentals, you should realize that you are not alone if you think you are an impostor—that it's a fluke when you happen to do well on a test, and that on the next test, *for sure* they (and your family and friends) are finally going to figure out how incompetent you really are. This feeling is so extraordinarily common that it even has a name—the "impostor phenomenon."[15] If you suffer from these kinds of feelings of inadequacy, just be aware that many others secretly share them.

Everyone has different gifts. As the old saying goes, "When one door closes, another opens." Keep your chin up and your eye on the open door.

REACHING TOWARD THE INFINITE

Some feel that diffuse, intuitive ways of thinking are more in tune with our spirituality. The creativity that diffuse thinking promotes sometimes seems beyond human understanding.

As Albert Einstein noted, "There are only two ways to live your life. One is as though nothing is a miracle. The other is as if everything is."

DON'T UNDERESTIMATE YOURSELF

"I coach Science Olympiad at our school. We have won the state championship eight out of the last nine years. We fell one point short of winning the state this year, and we often finish in the top ten in the nation. We have found that many seemingly top students (who are getting an A+ in all their classes) do not perform as well under the pressure of a Science Olympiad event as those who can mentally manipulate the knowledge they have. Interestingly, this second tier (if you will) of students at times seem to think of themselves as less intelligent than these top students. I would much rather take ostensibly lower-performing students who can think creatively on their feet, as the Olympiad requires, than top students who get flustered if the questions being posed don't exactly fit the memorized chunks in their brains."

—*Mark Porter, biology teacher, Mira Loma High School,*
Sacramento, California

SUMMING IT UP

- At some point, after you've got chunked material well in hand (and in brain), you start to let go of conscious awareness of every little detail and do things automatically.
- It may seem intimidating to work alongside other students who grasp material more quickly than you do. But "average" students can sometimes have advantages when it comes to initiative, ability to get things done, and creativity.
- Part of the key to creativity is to be able to switch from full focused concentration to the relaxed, daydreamy diffuse mode.
- Focusing too intently can *inhibit* the solution you are seeking—like trying to hammer a screw because you think it's a nail. When you are stuck, sometimes it's best to get away from a problem for a while and move on to something else, or to simply sleep on it.

PAUSE AND RECALL

Close the book and look away. What were the main ideas of this chapter? Pause also to try to recall the essential ideas of the book as a whole so far.

ENHANCE YOUR LEARNING

1. Think of an area where persistence has paid off for you in your life. Is there a new area where you would like to start developing your persistence? What backup plan can you develop for low times when you might feel like faltering?

2. People often try to stop their daydreaming, because it interrupts activities they truly intend to focus on, like listening to an important lecture. What works better for you—forcing yourself to maintain focus, or simply bringing your attention back to the matter at hand when you notice your attention wandering?

FROM SLOW LEARNER TO SUPERSTAR: NICK APPLEYARD'S STORY

Nick Appleyard leads the Americas business unit as a vice president in a high-tech company that develops and supports advanced physics simulation tools used in aerospace, automotive, energy, biomedical, and many other sectors of the economy. He received his degree in mechanical engineering degree from the University of Sheffield in England.

"Growing up, I was branded a slow learner and a problem child because of it. These labels impacted me deeply. I felt like my teachers treated me as if they'd given up any hope that I could succeed. To make matters worse, my parents also became frustrated with me and my educational progress. I felt the disappointment most severely from my father, a senior physician at a major teaching hospital. (I learned later in life that he had had similar difficulties early in his childhood.) It was a vicious circle that impacted my confidence in every aspect of life.

"What was the problem? Math and everything associated with it—fractions, times tables, long division, algebra, you name it. It was all boring and completely pointless.

"One day, something began to change, although I didn't realize it at the time. My father brought home a computer. I had heard about kids in their teens writing home computer games that everyone wanted to play, and becoming millionaires overnight. I wanted to be one of those kids.

"I read, practiced, and wrote harder and harder programs, all of which involved some kind of math. Eventually, a popular UK computer magazine accepted one of my programs for publication—a real thrill for me.

"Now I see every day how mathematics is applied for designing the next generation of automobiles, for helping to put rockets into space, and for analyzing how the human body works.

"Mathematics is no longer pointless. It is instead a source of wonder—and of a great career!"

{ 13 }

sculpting your brain

This time, eleven-year-old Santiago Ramón y Cajal's crime had been to build a small cannon and blow a neighbor's new, large wooden gate into splinters. In rural Spain of the 1860s, there weren't many options for oddball juvenile delinquents. That's how the young Cajal found himself locked in a flea-ridden jail.

Cajal was stubborn and rebellious. He had a single overwhelming passion: art. But what could he do with painting and drawing? Especially since Cajal ignored the rest of his studies—particularly math and science, which he thought were useless.

Cajal's father, Don Justo, was a strict man who had brought himself up from virtually nothing. The family was definitely not on aristocratic easy street. To try and give his son much-needed discipline and stability, Don Justo apprenticed him out to a barber. This was a disaster, as Cajal just neglected his studies even further. Beaten and starved by his teachers in an attempt to bring him around, Cajal was a mocking, shocking disciplinary nightmare.

Santiago Ramón y Cajal won the Nobel Prize for his many important contributions to our understanding of the structure and function of the nervous system.[1] In this picture, Cajal looks more like an artist than a scientist. His eyes show a hint of the same mischief that brought him so much trouble as a child.

Cajal met and worked with many brilliant scientists through his lifetime, people who were often far smarter than he. In Cajal's revealing autobiography, however, he pointed out that although brilliant people can do exceptional work, just like anyone else, they can also be careless and biased. Cajal felt the key to his success was his perseverance (the "virtue of the less brilliant"[2]) coupled with his flexible ability to change his mind and admit errors. Underlying everything was the support of his loving wife, Doña Silvería Fañanás García (the couple had seven children). *Anyone*, Cajal noted, even people with average intelligence, can sculpt their own brain, so that even the least gifted can produce an abundant harvest.[3]

Who knew that Santiago Ramón y Cajal would one day not only earn the Nobel Prize, but eventually become known as the father of modern neuroscience?

Change Your Thoughts, Change Your Life

Santiago Ramón y Cajal was already in his early twenties when he began climbing from bad-boy delinquency into the traditional study of medicine. Cajal himself wondered if perhaps his head had simply "grown weary of frivolity and irregular behavior and was beginning to settle down."[4]

There's evidence that myelin sheaths, the fatty insulation that helps signals move more quickly along a neuron, often don't finish developing until people are in their twenties. This may explain why teenagers often have trouble controlling their impulsive behavior— the wiring between intention and control areas isn't completely formed.[5]

....

"Deficiencies of innate ability may be compensated for through persistent hard work and concentration. One might say that work substitutes for talent, or better yet that it *creates talent*."[6]

—Santiago Ramón y Cajal

....

When you *use* neural circuits, however, it seems you help build the myelin sheath over them—not to mention making many other microscopic changes.[7] Practice appears to strengthen and reinforce connections between different brain regions, creating highways between the brain's control centers and the centers that store knowledge. In Cajal's case, it seems his natural maturation processes, coupled with his own efforts to develop his thinking, helped him to take control of his overall behavior.[8]

It seems people can *enhance* the development of their neuronal circuits by practicing thoughts that *use* those neurons.[9] We're still in the infancy of understanding neural development, but one thing is becoming clear—**we can make significant changes in our brain by changing how we think.**

What's particularly interesting about Cajal is that he achieved his greatness even though he *wasn't* a genius—at least, not in the conventional sense of the term. Cajal deeply regretted that he never had a "quickness, certainty, and clearness in the use of words."[10] What's worse is that when Cajal got emotional, he lost his way with words almost entirely. He couldn't remember things by rote, which made school, where parroting back information was prized, agony for him. The best Cajal could do was to grasp and remember key ideas; he frequently despaired his modest powers of understanding.[11] Yet some of the most exciting areas of neuroscientific research today are rooted in Cajal's original findings.[12]

Cajal's teachers, as Cajal later recollected, showed a sadly mistaken valuing of abilities. Quickness was taken as cleverness, memory for ability, and submissiveness for rightness.[12] Cajal's success despite his "flaws" shows us how even today, teachers can easily underestimate their students—and students can underestimate themselves.

Deep Chunking

Cajal worked his way fitfully through medical school. After adventures in Cuba as an army doctor and several failed attempts at competitive examinations to place as a professor, he finally obtained a position as a professor of histology, studying the microscopic anatomy of biological cells.

Each morning in his work in studying the cells of the brain and the nervous system, Cajal carefully prepared his microscope slides.

Then he spent hours carefully viewing the cells that his stains had highlighted. In the afternoon, Cajal looked to the abstract picture of his mind's eye—what he could remember from his morning's viewings—and began to draw the cells. Once finished, Cajal compared his drawing with the image he saw in the microscope. Then Cajal went back to the drawing board and started again, redrawing, checking, and redrawing. Only after his drawing captured the synthesized essence, not of just a single slide, but of the entire collection of slides devoted to a particular type of cell, did Cajal rest.[14]

Cajal was a master photographer—he was even the first to write a book in Spanish on how to do color photography. But he never felt that photographs could capture the true *essence* of what he was seeing. Cajal could only do that through his art, which helped him abstract—*chunk*—reality in a way that was most useful for helping others see the essence of the chunks.

A synthesis—an abstraction, chunk, or gist idea—is a neural pattern. **Good chunks form neural patterns that resonate, not only within the subject we're working in, but with other subjects and areas of our lives. The abstraction helps you transfer ideas from one area to another.**[15] That's why great art, poetry, music, and literature can be so compelling. When we grasp the chunk, it takes on a new life in our own minds—we form ideas that enhance and enlighten the neural patterns we already possess, allowing us to more readily see and develop other related patterns.

Once we have created a chunk as a neural pattern, we can more easily pass that chunked pattern to others, as Cajal and other great artists, poets, scientists, and writers have done for millennia. Once other people grasp that chunk, not only can they use it, but also they can more easily create similar chunks that apply to other areas in their lives—an important part of the creative process.

Here you can see that the chunk—the rippling neural ribbon—on the left is very similar to the chunk on the right. This symbolizes the idea that once you grasp a chunk in one subject, it is much easier for you to grasp or create a similar chunk in another subject. The same underlying mathematics, for example, echo throughout physics, chemistry, and engineering—and can sometimes also be seen in economics, business, and models of human behavior. This is why it can be easier for a physics or engineering major to earn a master's in business administration than someone with a background in English or history.[16]

Metaphors and physical analogies also form chunks that can allow ideas even from very different areas to influence one another.[17] This is why people who love math, science, and technology often also find surprising help from their activities or knowledge of sports, music, language, art, or literature. My own knowledge of how to learn a language helped me in learning how to learn math and science.

One important key to learning swiftly in math and science is to realize that virtually every concept you learn has an analogy—a comparison—with something you already know.[18] Sometimes the analogy or metaphor is rough—such as the idea that blood vessels are like highways, or that a nuclear reaction is like falling dominoes. But these simple analogies and metaphors can be powerful tools to help you use an existing neural structure as a scaffold to help you more rapidly build a new, more complex neural structure. As you begin to use this new structure, you will discover that it has features that make it far more useful than your first simplistic structure. These new structures can in turn become sources of metaphor and analogy for still newer ideas in very different areas. (This, indeed, is why physicists and engineers have been sought after in the world of finance.) Physicist Emanual Derman, for example, who did brilliant research in theoretical particle physics, moved on to the company Goldman Sachs, eventually helping to develop the Black-Derman-Toy interest-rate model. Derman eventually took charge of the firm's Quantitative Risk Strategies group.

SUMMING IT UP

- Brains mature at different speeds. Many people do not develop maturity until their midtwenties.
- Some of the most formidable heavyweights in science started out as apparently hopeless juvenile delinquents.
- One trait that successful professionals in science, math, and technology gradually learn is how to chunk—to abstract key ideas.
- Metaphors and physical analogies form chunks that can allow ideas from very different areas to influence one another.
- Regardless of your current or intended career path, keep your mind open and ensure that math and science are in your learning repertoire. This gives you a rich reserve of chunks to help you be smarter about your approach to all sorts of life and career challenges.

PAUSE AND RECALL

Close the book and look away. What were the main ideas of this chapter? You will find that you can recall these ideas more easily if you relate them to your own life and career goals.

ENHANCE YOUR LEARNING

1. In his career, Santiago Ramón y Cajal found a way to combine his passion for art with a passion for science. Do you know other people, either famous public figures or family friends or acquaintances, who have done something similar? Is such a confluence possible in your own life?

2. How can you avoid falling into the trap of thinking that quicker people are automatically more clever?

3. Doing what you are told to do can have benefits and drawbacks. Compare Cajal's life with your own. When has doing what you were told been beneficial? When has it inadvertently created problems?

4. Compared to Cajal's handicaps, how do your own limitations stack up? Can you find ways to turn your disadvantages into advantages?

{ 14 }

developing the mind's eye through equation poems

Learn to Write an Equation Poem—
Unfolding Lines That Provide a Sense
of What Lies Beneath a Standard Equation

Poet Sylvia Plath once wrote: "The day I went into physics class it was death."[1] She continued:

> A short dark man with a high, lisping voice, named Mr. Manzi, stood in front of the class in a tight blue suit holding a little wooden ball. He put the ball on a steep grooved slide and let it run down to the bottom. Then he started talking about let a equal acceleration and let t equal time and suddenly he was scribbling letters and numbers and equals signs all over the blackboard and my mind went dead.

Mr. Manzi had, at least in this semiautobiographical retelling of Plath's life, written a four-hundred-page book with no drawings or photographs, only diagrams and formulas. An equivalent would be trying to appreciate Plath's poetry by being *told* about it, rather than being able to read it for yourself. Plath was, in her version of the story, the only student to get an A, but she was left with a dread for physics.

> "What, after all, is mathematics but the poetry of the mind, and what is poetry but the mathematics of the heart?"
>
> —*David Eugene Smith, American mathematician and educator*

Physicist Richard Feynman's introductory physics classes were entirely different. Feynman, a Nobel Prize winner, was an exuberant guy who played the bongos for fun and talked more like a down-to-earth taxi driver than a pointy-headed intellectual.

When Feynman was about eleven years old, an off-the-cuff remark had a transformative impact on him. He remarked to a friend that thinking is nothing more than talking to yourself inside.

"Oh yeah?" said Feynman's friend. "Do you know the crazy shape of the crankshaft in a car?"

"Yeah, what of it?"

"Good. Now tell me: How did you describe it when you were talking to yourself?"

It was then that Feynman realized that thoughts can be visual as well as verbal.[2]

He later wrote about how, when he was a student, he had struggled to imagine and visualize concepts such as electromagnetic waves, the invisible streams of energy that carry everything from sunlight to cell phone signals. He had difficulty describing what he saw in his mind's eye.[3] If even one of the world's greatest physicists had trouble imagining how to see some (admittedly difficult-to-imagine) physical concepts, where does that leave us normal folks?

We can find encouragement and inspiration in the realm of poetry.[4] Let's take a few poetic lines from a song by American singer-
songwriter Jonathan Coulton, called "Mandelbrot Set,"[5] about a
famous mathematician, Benoit Mandelbrot.

> *Mandelbrot's in heaven*
> *He gave us order out of chaos, he gave us hope where there was none*
> *His geometry succeeds where others fail*
> *So if you ever lose your way, a butterfly will flap its wings*
> *From a million miles away, a little miracle will come to take you home*

The essence of Mandelbrot's extraordinary mathematics is captured
in Coulton's emotionally resonant phrases, which form images that
we can see in our own mind's eye—the gentle flap of a butterfly's
wings that spreads and has effects even a million miles away.

Mandelbrot's work in creating a new geometry allowed us to understand that sometimes, things that look rough and messy—like
clouds and shorelines—have a degree of order to them. Visual complexity can be created from simple rules, as evidenced in modern
animated movie-making magic. Coulton's poetry also alludes to the
idea, embedded in Mandelbrot's work, that tiny, subtle shifts in one
part of the universe ultimately affect everything else.

The more you examine Coulton's words, the more ways you can
see it applied to various aspects of life—these meanings become
clearer the more you know and understand Mandelbrot's work.

**There are hidden meanings in equations, just as there are in
poetry.** If you are a novice looking at an equation in physics, and
you're not taught how to see the life underlying the symbols, the lines
will look dead to you. It is when you begin to learn and supply the
hidden text that the meaning slips, slides, then finally leaps to life.

In a classic paper, physicist Jeffrey Prentis compares how a
brand-new student of physics and a mature physicist look at equa-

tions.[6] The equation is seen by the novice as just one more thing to memorize in a vast collection of unrelated equations. More advanced students and physicists, however, see with their mind's eye the *meaning* beneath the equation, including how it fits into the big picture, and even a sense of how the parts of the equation *feel*.

....

"A mathematician who is not at the same time something of a poet will never be a full mathematician."

—*German mathematician Karl Weierstrass*

....

When you see the letter *a*, for acceleration, you might feel a sense of pressing on the accelerator in a car. Zounds! *Feel* the car's acceleration pressing you back against the seat.

Do you need to bring these feelings to mind every time you look at the letter *a*? Of course not; you don't want to drive yourself crazy remembering every little detail underlying your learning. But that sense of pressing acceleration should hover as a chunk in the back of your mind, ready to slip into working memory if you're trying to analyze the meaning of *a* when you see it roaming around in an equation.

Similarly, when you see *m*, for *mass*, you might feel the inertial laziness of a fifty-pound boulder—it takes a lot to get it moving. When you see the letter *f*, for *force*, you might see with your mind's eye what lies underneath *force*—that it depends on both *mass* and *acceleration*: $m \cdot a$, as in the equation $f = m \cdot a$. Perhaps you can feel what's behind the *f* as well. Force has built into it a heaving *oomph* (acceleration), against the lazy *mass* of the boulder.

Let's build on that just a wee bit more. The term *work* in physics means energy. We do *work* (that is, we supply energy) when we push (*force*) something through a *distance*. We can encrypt that with poetic simplicity: $w = f \times d$. Once we see w for work, then we can imagine with our mind's eye, and even our body's feelings, what's behind it. Ultimately, we can distill a line of equation poetry that looks like this:

$$w$$
$$w = f \cdot d$$
$$w = (ma) \cdot d$$

Symbols and equations, in other words, have a hidden text that lies beneath them—a meaning that becomes clear once you are more familiar with the ideas. Although they may not phrase it this way, scientists often see equations as a form of poetry, a shorthand way to symbolize what they are trying to see and understand. Observant people recognize the depth of a piece of poetry—it can have many possible meanings. In just the same way, maturing students gradually learn to see the hidden meaning of an equation with their mind's eye and even to intuit different interpretations. It's no surprise to learn that graphs, tables, and other visuals also contain hidden meaning—meaning that can be even more richly represented in the mind's eye than on the page.

Simplify and Personalize Whatever You Are Studying

We've alluded to this before, but it's worth revisiting now that we've got better insight into how to imagine the ideas that underlie equations. **One of the most important things we can do when we are trying to learn math and science is to bring the abstract ideas to life in our minds.** Santiago Ramón y Cajal, for example, treated the microscopic scenes before him as if they were inhabited by living creatures that hoped and dreamed just as people themselves do.[7] Cajal's colleague and friend, Sir Charles Sherrington, who coined

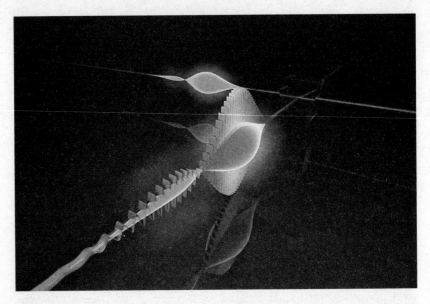

Einstein was able to imagine himself as a photon.[8] We can gain a sense of what Einstein saw by looking at this beautiful vision by Italian physicist Marco Bellini of an intense laser pulse (the one in front), being used to measure the shape of a single photon (the one in the back).

the word *synapse,* told friends that he had never met another scientist who had this intense ability to breathe life into his work. Sherrington wondered whether this might have been a key contributing factor to Cajal's level of success.

Einstein's theories of relativity arose not from his mathematical skills (he often needed to collaborate with mathematicians to make progress) but from his ability to pretend. He imagined himself as a photon moving at the speed of light, then imagined how a second photon might perceive him. What would that second photon see and feel?

Barbara McClintock, who won the Nobel Prize for her discovery of genetic transposition ("jumping genes" that can change their place on the DNA strand), wrote about how she imagined the corn plants she studied: "I even was able to see the internal parts of the

Pioneering geneticist Barbara McClintock imagined gigantic versions of the molecular elements she was dealing with. Like other Nobel Prize winners, she personalized—even made friends with—the elements she was studying.

chromosomes—actually everything was there. It surprised me because I actually felt as if I were right down there and these were my friends."[9]

It may seem silly to stage a play in your mind's eye and imagine the elements and mechanisms you are studying as living creatures, with their own feelings and thoughts. But it is a method that works—it brings them to life and helps you see and understand phenomena that you couldn't intuit when looking at dry numbers and formulas.

Simplifying is also important. Richard Feynman, the bongo-playing physicist we met earlier in this chapter, was famous for asking scientists and mathematicians to explain their ideas in a simple way so that he could grasp them. Surprisingly, simple explanations are possible for almost *any* concept, no matter how complex. When you cultivate simple explanations by breaking down complicated material to its key elements, the result is that you have a deeper understanding of the material.[10] Learning expert Scott Young has developed this idea in what he calls the *Feynman technique*, which asks people to find a simple metaphor or analogy to help them grasp the essence of an idea.[11]

The legendary Charles Darwin would do much the same thing. When trying to explain a concept, he imagined someone had just walked into his study. He would put his pen down and try to explain the idea in the simplest terms. That helped him figure out how he would describe the concept in print. Along those lines, the website Reddit.com has a section called "Explain like I'm 5" where anyone can make a post asking for a simple explanation of a complex topic.[12]

You may think you really have to understand something in order to explain it. But observe what happens when you are talking to other people about what you are studying. You'll be surprised to see how often understanding arises as a *consequence* of attempts to explain to others and yourself, rather than the explanation arising out of your previous understanding. This is why teachers often say that the first time they ever really understood the material was when they had to teach it.

IT'S NICE TO GET TO KNOW YOU!

"Learning organic chemistry is not any more challenging than getting to know some new characters. The elements each have their own unique personalities. The more you understand those personalities, the more you will be able to read their situations and predict the outcomes of reactions."

—*Kathleen Nolta, Ph.D., Senior Lecturer in Chemistry and recipient of the Golden Apple Award, recognizing excellence in teaching at the University of Michigan*

NOW YOU TRY!

Stage a Mental Play

Imagine yourself within the realm of something you are studying—looking at the world from the perspective of the cell, or the electron, or even a mathematical concept. Try staging a mental play with your new friends, imagining how they feel and react.

Transfer—Applying What You've Learned in New Contexts

Transfer is the ability to take what you learn in one context and apply it to something else. For example, you may learn one foreign language and then find that you can pick up a second foreign language more easily than the first. That's because when you learned the first foreign language, you also acquired general language-learning skills, and potentially similar new words and grammatical structures, that *transferred* to your learning the second foreign language.[13]

Learning math by applying it only to problems within a specific discipline, such as accounting, engineering, or economics, can be a little like deciding that you are not really going to learn a foreign language after all—you're just going to stick to one language and just learn a few extra English vocabulary words. Many mathematicians feel that learning math through entirely discipline-specific approaches makes it more difficult for you to use mathematics in a flexible and creative way.

Mathematicians feel that if you learn math the way they teach it, which centers on the abstract, chunked essence without a specific application in mind, you've captured skills that are easy for you to transfer to a variety of applications. In other words, you'll have acquired the equivalent of general language-learning skills. You may be a physics student, for example, but you could use your knowledge of abstract math to quickly grasp how some of that math could apply to very different biological, financial, or even psychological processes.

This is part of why mathematicians like to teach math in an abstract way, without necessarily zooming in on applications. They want you to see the essence of the ideas, which they feel makes it easier to transfer the ideas to a variety of topics.[14] It's as if they don't want you to learn how to say a specific Albanian or Lithuanian or Icelandic phrase meaning *I run* but rather to understand the more general idea that there is a category of words called *verbs*, which you *conjugate*.

The challenge is that it's often easier to pick up on a mathematical idea if it is applied directly to a concrete problem—even though that can make it more difficult to transfer the mathematical idea to new areas later. Unsurprisingly, there ends up being a constant tussle between concrete and abstract approaches to learning mathematics. Mathematicians try to hold the high ground by stepping back to make sure that abstract approaches are central to the learning process. In contrast, engineering, business, and many other professions all naturally gravitate toward math that focuses on their specific areas to help build student engagement and avoid the complaint of "When am I ever going to use this?" Concretely applied math also gets around the issue that many "real-world" word problems in mathematics textbooks are simply thinly disguised exercises. In the end, both concrete and abstract approaches have their advantages and disadvantages.

Transfer is beneficial in that it often makes learning easier for students as they advance in their studies of a discipline. As Professor Jason Dechant of the University of Pittsburgh says, "I always tell my students that they will study less as they progress through their nursing programs, and they don't believe me. They're actually doing more and more each semester; they just get better at bringing it all together."

One of the most problematic aspects of procrastination—constantly interrupting your focus to check your phone messages, e-mails, or other updates—is that it interferes with transfer. Students who interrupt their work constantly not only don't learn as deeply, but also aren't able to transfer what little they do learn as easily to other topics.[15] You may think you're learning in between checking your phone messages, but in reality, your brain is not focusing long enough to form the solid neural chunks that are central to transferring ideas from one area to another.

TRANSFERRING IDEAS *WORKS!*

"I took fishing techniques from the Great Lakes and tried using them down in the Florida Keys this past year. Completely different fish, different bait, and a technique that had never been used but it worked great. People thought I was crazy and it was funny to show them that it actually caught fish."

—*Patrick Scoggin, senior, history*

SUMMING IT UP

- Equations are just ways of abstracting and simplifying concepts. This means that equations contain deeper meaning, similar to the depth of meaning found in poetry.
- Your "mind's eye" is important because it can help you stage plays and personalize what you are learning about.
- Transfer is the ability to take what you learn in one context and apply it to something else.

- It's important to grasp the chunked essence of a mathematical concept, because then it's easier to transfer and apply that idea in new and different ways.
- Multitasking during the learning process means you don't learn as deeply—this can inhibit your ability to transfer what you are learning.

PAUSE AND RECALL

Close the book and look away. What were the main ideas of this chapter? Can you picture some of these ideas with symbols in your mind's eye?

ENHANCE YOUR LEARNING

1. Write an equation poem—several unfolding lines that provide a sense of what lies beneath a standard equation.

2. Write a paragraph that describes how some concepts you are studying could be visualized in a play. How do you think the actors in your play might realistically feel and react to one another?

3. Take a mathematical concept you have learned and look at a concrete example of how that concept is applied. Then step back and see if you can sense the abstract chunk of an idea underlying the application. Can you think of a completely different way that concept might be used?

{ 15 }

renaissance learning

The Value of Learning on Your Own

People like Charles Darwin, whose theory of evolution has made him one of the most influential figures in human history, are often thought of as natural geniuses. You may be surprised to learn that much like Cajal, Darwin was a poor student. He washed out of medical school and ended up, to his father's horror, heading out on a round-the-world voyage as the ship's naturalist. Out on his own, Darwin was able to look with fresh eyes at the data he was collecting.

Persistence is often more important than intelligence.[1] Approaching material with a goal of learning it on your own gives you a unique path to mastery. Often, no matter how good your teacher and textbook are, it's only when you sneak off and look at other books or videos that you begin to see that what you learn through a single teacher or book is a partial version of the full, three-dimensional reality of the subject, which has links to still *other* fascinating topics that are of *your* choosing.

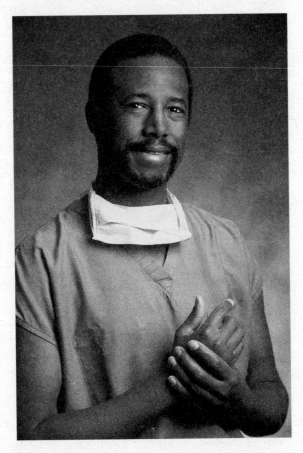

Neurosurgeon Ben Carson, winner of the Presidential Medal of Freedom for his pioneering surgical innovations, was initially flunking and gently urged to leave medical school. Carson knew he learned best through books, not in-class lectures. He took a counterintuitive step and *stopped* attending lectures to give himself time to focus on learning through books. His grades soared and the rest is history. (Note that this technique would not work for everyone—and if you use this story as an excuse to simply stop attending classes, you are courting disaster!)

In the fields of science, math, and technology, many individuals had to carve their own path in learning, either because they had no other way, or because for whatever reason, they'd thrown away previous learning opportunities. Research has shown that students learn best when they themselves are actively engaged in the subject instead of simply listening to someone else speak.[2] A student's ability to grapple personally with the material, sometimes bouncing it off fellow learners, is key.

Santiago Ramón y Cajal was horrified when he had to learn college calculus as an adult, after he had become serious about becoming a doctor. He'd never paid attention to math in his youth and lacked even a rudimentary understanding of the material. He had to go rummaging back through old books, scratching his head to figure out the basics. Cajal learned all the more deeply, however, because he was driven by his personal goals.

....

"What a wonderful stimulant it would be for the beginner if his instructor, instead of amazing and dismaying him with the sublimity of great past achievements, would reveal instead the origin of each scientific discovery, the series of errors and missteps that preceded it—information that, from a human perspective, is essential to an accurate explanation of the discovery."[3]

—Santiago Ramón y Cajal

....

Inventor and author William Kamkwamba, born in 1987 in Africa, could not afford to attend school. So he began teaching himself by going to his village's library, where he stumbled across a book titled *Using Energy*. But Kamkwamba didn't just read the book. When he was only fifteen years old, he used the book to guide him in active learning: He built his own windmill. His neighbors called him *misala*—crazy—but his creation helped begin generating electricity and running water for his village and sparked the growth of grassroots technological innovation in Africa.[4]

American neuroscientist and pharmacologist Candace Pert had

an excellent education, earning a doctorate in pharmacology from Johns Hopkins University. But part of her inspiration and subsequent success arose from an unusual source. Just before entering medical graduate school, she hurt her back in a horseback-riding accident and spent a summer under the influence of deep pain medication.[5] Her personal experiences with pain and pain medication drove her scientific research. Ignoring her advisor's attempts to stop her, she made some of the first key discoveries involving opiate receptors—a major step forward in understanding addiction.

College isn't the only way to learn. Some of the most powerful and renowned people of our time, including powerhouses Bill Gates, Larry Ellison, Michael Dell, Mark Zuckerberg, James Cameron, Steve Jobs, and Steve Wozniak, dropped out of college. We will continue to see fascinating innovations from people who are able to combine the best aspects of traditional and nontraditional learning with their own self-taught approaches.

Taking responsibility for your own learning is one of the most important things you can do. Teacher-centered approaches, where the teacher is considered to be the one with the answers, may sometimes inadvertently foster a sense of helplessness about learning among students.[6] Surprisingly, teacher evaluation systems may foster the same helplessness—these systems allow you to place the blame for failure on your teacher's inability to motivate or instruct.[7] Student-centered learning, where students are challenged to learn from one another and are expected to be their own drivers toward mastery of the material, is extraordinarily powerful.

The Value of Great Teachers

You will also sometimes have a chance to interact with truly special mentors or teachers. When this lucky opportunity arises, seize it.

Train yourself to get past the *gulp* stage and force yourself to reach out and ask questions—real and to-the-point questions, not questions meant to show off what you know. The more you do this, the easier it will become, and the more helpful it will be in ways you never anticipated—a simple sentence growing from their vast experience can change the course of your future. And also be sure to show appreciation for the people guiding you—it is essential to let them know that the help is meaningful.

Be wary, however, of falling into "sticky student" syndrome. Kind teachers, in particular, can become magnets for students whose true needs involve desire for the ego-boosting attention of the instructor far more than answers to the actual questions being posed. Well-meaning teachers can burn out trying to satisfy never-to-be-satisfied desires.

Also avoid the trap of feeling *certain* your answer is correct, and attempting to force your teacher to follow the tortured steps of your logic when your answer is obviously wrong. Every once in a while, you might ultimately be proven correct, but for many teachers, particularly at more advanced levels in math and science, trying to follow twisted, erroneous thinking is like listening to out-of-tune music—a thankless, painful exercise. It's generally best to start your thinking afresh and listen to your teacher's suggestions. When you finally understand the answer, you can go back if you want to debug your previous error. (Often you'll realize, in one fell swoop, that it's difficult to even put into words how wrong your previous approach was.) Good teachers and mentors are often very busy people, and you need to use their time wisely.

Truly great teachers make the material seem both simple and profound, set up mechanisms for students to learn from each other, and inspire students to learn on their own. Celso Batalha, for example, a renowned professor of physics at Evergreen Valley College, has set up a popular reading group for his students about learning how to learn. And many professors use "active" and "collaborative teaching" techniques in the classroom that give students a chance to actively engage with the material and with each other.[8]

One thing has surprised over the years. Some of the greatest teachers I've ever met told me that when they were young, they were too shy, too tongue-tied in front of audiences, and too intellectually incapable to ever dream of becoming a teacher. They were ultimately surprised to discover that the qualities they saw as disadvantages helped propel them into being the thoughtful, attentive, creative instructors and professors they became. It seemed their introversion made them more thoughtful and sensitive to others, and their humble awareness of their past failings gave them patience and kept them from becoming aloof know-it-alls.

The *Other* Reason for Learning on Your Own— Quirky Test Questions

Let's return to the world of traditional learning in high school and college, where a little insider knowledge will help you succeed. One secret of math and science teachers is that they often take quiz and test questions from books that *aren't* in the assigned reading for the course. After all, it's hard to come up with new test questions each semester. This means that test questions often have slight differences in terminology or approach that can throw you off your game even if you are comfortable with your textbook and your teacher's lectures. You can end up thinking that you don't have talent for math and science, when all you really needed to do was look at the material through different lenses as you were studying throughout the semester.

Beware of Intellectual Snipers

Santiago Ramón y Cajal had a deep understanding not only of how to conduct science, but also of how people interact with one another. He warned fellow learners that *there will always be those who criticize or attempt to undermine any effort or achievement you make.* This happens to everyone, not just Nobel Prize winners. If you do well in your studies, the people around you can feel threatened. The greater your achievement, the more other people will sometimes attack and demean your efforts.

On the other hand, if you flunk a test, you may also encounter critics who throw more barbs, saying you don't have what it takes. Failure is not so terrible. Analyze what you did wrong and use it to correct yourself to do better in the future. Failures are better teachers than successes because they cause you to rethink your approach.

Some "slower" students struggle with math and science because they can't seem to understand ideas that others find obvious. These students unfortunately sometimes think of themselves as not very bright, but the reality is that their slower way of thinking can allow them to see confusing subtleties that others aren't aware of. It's the equivalent of a hiker who notices the scent of pine and small-animal paths in the woods, as opposed to the oblivious motorist who's whizzing by at seventy miles an hour. Sadly, some instructors feel threatened by the deceptively simple questions that seemingly pedestrian students can pose. Rather than acknowledging how perceptive these questions are, instructors attack the questioner with brusque, brush-off answers that equate to "just do as you're told like everybody else does." This leaves the questioner feeling foolish and only deepens the confusion. (Keep in mind that instructors sometimes can't tell whether you are thinking deeply about the material, or whether you're having trouble taking ownership of your own role in understanding simple matters, as was the case with my truculent behavior in high school.)

In any case, if you find yourself struggling with the "obvious," don't despair. Look to your classmates or the Internet for help. One useful trick is to try to find another instructor—one with nice evaluations—who occasionally teaches the same class. These instructors often understand what you are experiencing and are sometimes willing to help if you don't overuse them as a resource. Remind yourself that this situation is only temporary, and no circumstance is truly as overwhelming as it might seem at the time.

As you will find when you reach the work world (if you haven't already), many individuals are far more interested in affirming their own ideas and making themselves look good than they are in helping you. In this kind of situation, there can be a fine line between keeping yourself open to constructive explanation and criticism, versus being closed to commentary or criticism that is couched as constructive but is actually simply spiteful. Whatever the criticism, if you feel a strong wash of emotion or certainty ("But I'm right!"), it may be a clue that you're correct—or alternatively (and perhaps even more likely, because of your telltale emotions), it may be that you need to go back and reexamine matters using a more objective perspective.

We're often told that empathy is universally beneficial, but it's not.[9] It's important to learn to switch on an occasional cool dispassion that helps you to not only focus on what you are trying to learn, but also to tune people out if you discover their interests lie in undercutting you. Such undercutting is all-too-common, as people are often just as competitive as they are cooperative. When you're a young person, mastering such dispassion can be difficult. We are naturally excited about what we are working on, and we like to believe that everyone can be reasoned with and that almost everyone is naturally good-hearted toward us.

Like Cajal, you can take pride in aiming for success *because of* the very things that make other people say you can't do it. **Take pride in who you are, especially in the qualities that make you "different," and use them as a secret talisman for success.** Use your natural contrariness to defy the always-present prejudices from others about what you can do.

NOW YOU TRY!

Understanding the Value of "Bad"

Pick a seemingly bad trait and describe how it could be beneficial in helping you learn or think creatively or independently. Could you think of a way to diminish the negative aspects of that trait, even as you enhance the positive aspects?

SUMMING IT UP

- Learning on your own is one of the deepest, most effective ways to approach learning:
  - It improves your ability to think independently.
  - It can help you answer the strange questions that teachers sometimes throw at you on tests.
- In learning, persistence is often far more important than intelligence.
- Train yourself to occasionally reach out to people you admire. You can gain wise new mentors who, with a simple sentence, can change the course of your future. But use your teachers' and mentors' time sparingly.
- If you aren't very fast at grasping the essentials of whatever you are studying, don't despair. Surprisingly often, "slower" students are grappling with fundamentally important is-

sues that quicker students miss. When you finally get what's going on, you can get it at a deeper level.

- People are competitive as well as cooperative. There will always be those who criticize or attempt to undermine any effort or achievement you make. Learn to deal dispassionately with these issues.

{

PAUSE AND RECALL

Close the book and look away. What were the main ideas of this chapter? Which idea is most important— or are there several competing equally important ideas?

ENHANCE YOUR LEARNING

1. What are the advantages and disadvantages of learning on your own, without being guided by a formal program of study?

2. Look up the phrase *List of autodidacts* on Wikipedia. Which of the many autodidacts there would you most like to emulate? Why?

3. Choose a person among your own acquaintances (that is, not a celebrity) whom you admire but to whom you have never really spoken. Formulate a plan to say hello and introduce yourself—then carry it out.

NEW YORK TIMES SCIENCE WRITER
NICHOLAS WADE ON AN INDEPENDENT MIND

Nicholas Wade writes for the Science Times section of the *New York Times*. Always an independent thinker, Wade owes his very existence to the similar independent thinking of his grandfather—one of the few male survivors of the *Titanic*. When most men followed a rumor and moved to the port side, Wade's grandfather followed his intuition and deliberately moved the other way, to starboard.

Here, Nicholas gives his insight on what he thinks are the most interesting books about scientists and mathematicians.

"The Man Who Knew Infinity: A Life of the Genius Ramanujan, *by Robert Kanigel. This book tells the unbelievable, rags to intellectual riches story of the Indian mathematical genius Srinivasa Ramanujan and his friend British mathematician G. H. Hardy. My favorite episode is this:*

'Once, in the taxi from London, Hardy noticed its number, 1729. He must have thought about it a little because he entered the room where Ramanujan lay in bed and, with scarcely a hello, blurted out his disappointment with it. It was, he declared, "rather a dull number," adding that he hoped that wasn't a bad omen.

'"No, Hardy," said Ramanujan. "It is a very interesting number. It is the smallest number expressible as the sum of two cubes in different ways."'

"Noble Savages, *by Napoleon Chagnon. This beautifully written adventure story gives a sense of what it's like to learn to survive and thrive in an utterly alien culture. Chagnon was originally trained as an engineer. His scientific research has shifted our understanding of how cultures develop.*

"Men of Mathematics, *by E. T. Bell. This is an old classic that's a showstopping read for anyone who's interested in how fascinating people* think. Who could forget brilliant, doomed Évariste Galois who spent the night before he knew he was to die 'feverishly dashing off his last will and testament, writing against time to glean a few of the great things in his teeming mind before the death which he foresaw could overtake him. Time after time he broke off to scribble in the margin "I have not time; I have not time," and passed on to the next frantically scrawled outline.' Truth be told, this is one of the few exciting stories that Professor Bell perhaps exaggerated, although Galois unquestionably spent that last evening putting the final polish on his life's work. But this brilliant book has inspired generations of both men and women."

{ 16 }

avoiding
overconfidence

The Power of Teamwork

Fred had a problem. He couldn't move his left hand. This wasn't surprising. While singing in the shower, Fred had suffered a nearly lethal right-hemisphere ischemic stroke a month before. The brain's right hemisphere controls the left side of the body, which was why Fred's left hand was now lifeless.

Fred's real problem, though, was worse. Even though he couldn't move his left hand, Fred insisted—*and truly believed*—that he could. Sometimes he would excuse the lack of motion by saying he was just too tired to lift a finger. Or he'd insist that his left hand *had* moved. It was just that people hadn't been watching. Fred would even covertly move his left hand with his right, and then loudly proclaim that his hand had moved on its own.

Fortunately, as the months went by, Fred's left hand gradually regained its function. Fred laughed with his doctor about how he'd tricked himself into believing that he could move his hand in the weeks immediately following the stroke; he spoke cheerfully about returning to his work as an accountant.

But there were signs that Fred wasn't returning to business as usual. He used to be a caring, considerate guy, but the new Fred was dogmatic and self-righteous.

There were other changes. Fred used to be a keen practical joker, but now he just nodded along without understanding the punch lines to others' jokes. Fred's skill at investing also evaporated, and his cautiousness was replaced by naive optimism and overconfidence.

Even worse, Fred seemed to have become emotionally tone-deaf. He tried to sell his wife's car without asking her permission and was surprised when she became upset. When their beloved old family dog died, Fred sat placidly eating popcorn, watching his wife and children cry as if it were a scene out of a movie.

What made these changes more difficult to understand was that Fred seemed to have retained his intelligence—even his formidable way with numbers. He could still quickly work up a business profit-and-loss statement and solve complex algebra problems. One interesting anomaly, however, was that if Fred made a mistake in his calculations, concluding something nonsensical, such as that a hot dog stand had a loss of nearly a billion dollars, it didn't bother him. There was no big-picture "click" that said, "Wait a minute, that answer doesn't make sense."

It turns out that Fred is a typical victim of "broad-perspective perceptual disorder of the right hemisphere."[1] Fred's stroke had incapacitated broad areas of the right hemisphere of his brain. He could still function, but only partially.

Although we need to be careful about faulty and superficial "left brain/right brain" assumptions, we also don't want to throw the baby out with the bathwater and ignore worthwhile research

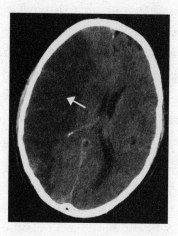

The arrow on this CT scan of the brain points toward the shadowed damage caused by a right-hemisphere ischemic stroke.

that gives intriguing hints about hemispheric differences.[2] Fred reminds us of the dangers of not using our full cognitive abilities, which involve many areas of our brain. Not using some of our abilities isn't as devastating for us as it is for Fred. But even subtle avoidance of some of our capabilities can have a surprisingly negative impact on our work.

Avoiding Overconfidence

There's a great deal of evidence from research that the right hemisphere helps us step back and put our work into big-picture perspective.[3] People with damage to the right hemisphere are often unable to gain "aha!" insights. That's why Fred wasn't able to catch the punch lines of jokes. The right hemisphere, it turns out, is vitally important in getting onto the right track and doing "reality checks."[4]

In some sense, **when you whiz through a homework or test problem and don't go back to check your work, you are acting a little like a person who is refusing to use parts of your brain.** You're not stopping to take a mental breath and then revisit what you've done with the bigger picture in mind, to see whether it makes sense.[5] As leading neuroscientist V. S. Ramachandran has noted, the right hemisphere serves as a sort of "'Devil's Advocate,' to question the status quo and look for global inconsistencies," while "the left hemisphere always tries to cling tenaciously to the way things were."[6] This echoes the pioneering work of psychologist Michael Gazzaniga, who posited that the left hemisphere interprets the world for us—and will go to great lengths to keep those interpretations unchanging.[7]

When you work in focused mode, it is easy to make minor mistakes in your assumptions or calculations. If you go off track early on, it doesn't matter if the rest of your work is correct—your answer is still wrong. Sometimes it's even laughably wrong—the equivalent of calculating a circumference of the earth that is only 2½ feet around. Yet these nonsensical results just don't matter to you, because the more left-centered focused mode has associated with it a desire to cling to what you've done.

That's the problem with the focused, left-hemisphere-leaning mode of analysis. It provides for an analytical and upbeat approach. But abundant research evidence suggests that there is a potential for rigidity, dogmatism, and egocentricity.

When you are absolutely certain that what you've done on a homework or test is fine—*thank you very much*—be aware that this feeling may be based on overly confident perspectives arising in part from the left hemisphere. When you step back and recheck, you are allowing for more interaction between hemispheres—taking advantage of the special perspectives and abilities of each.

People who haven't felt comfortable with math often fall into the trap of "equation sheet bingo." They desperately try to find a pattern in what the teacher or book did and fit their equations to that pattern. Good learners vet their work to ensure that it makes sense. They ask themselves what the equations mean and where they come from.

....

"The first principle is that you must not fool yourself—and you are the easiest person to fool."[8]

—*Physicist Richard Feynman, advising how to avoid pseudo-science that masquerades as science*

....

The Value of Brainstorming with Others

Niels Bohr was heavily involved in the Manhattan Project—the U.S. race during World War II to build the nuclear bomb before the Nazis. He was also one of the greatest physicists who ever lived— which ultimately made it difficult for him to think intelligently about physics.

Bohr was so respected as the genius who had intuited quantum theory that his thinking was considered unassailable. This meant that he could no longer brainstorm with others. No matter what cockamamie idea Bohr might propose, the other physicists work-ing on the bomb would *ooh* and *ahh* over it as if it were something sacred.

Bohr handled this challenge in an intriguing way.

Richard Feynman, as it turned out, was good at not being intimidated by other people—at simply doing physics, no matter who he was with. He was so good that he became Bohr's ace in the

Niels Bohr lounging with Albert Einstein in 1925.

hole. Feynman was at that time just a youngster in the crowd of hundreds of prominent physicists at Los Alamos. But he was singled out by Bohr to do private brainstorming together before Bohr would meet with the other physicists. Why? Feynman was the only one who wasn't intimidated by Bohr and who would tell Bohr that some of his ideas were foolish.[9]

As Bohr knew, brainstorming and working with others—as long

as they know the area—can be helpful. It's sometimes just not enough to use more of your own neural horsepower—both modes and hemispheres—to analyze your work. After all, everyone has blind spots. Your naively upbeat focused mode can still skip right over errors, especially if *you're* the one who committed the original errors.[10] Worse yet, sometimes you can blindly believe you've got everything nailed down intellectually, but you haven't. (This is the kind of thing that can leave you in shock when you discover you've flunked a test you'd thought you aced.)

By making it a point to do some of your studying with friends, you can more easily catch where your thinking has gone astray. Friends and teammates can serve as a sort of ever-questioning, larger-scale diffuse mode, outside your own brain, that can catch what you missed, or what you just can't see. And of course, as mentioned earlier, explaining to friends helps build your own understanding.

The importance of working with others doesn't just relate to problem solving—it's also important in career building. A single small tip from a teammate to take a course from the outstanding Professor Passionate, or to check out a new job opening, can make an extraordinary difference in how your life unfolds. One of the most-cited papers in sociology, "The Strength of Weak Ties," by sociologist Mark Granovetter, describes how the number of acquaintances you have—*not* the number of good friends—predicts your access to the latest ideas as well as your success on the job market.[11] Your good friends, after all, tend to run in the same social circles that you do. But acquaintances such as class teammates tend to run in different circles—meaning that your access to the "outside your brain" interpersonal diffuse mode is exponentially larger.

Those you study with should have, at least on occasion, an aggressively critical edge to them. Research on creativity in teams has shown that nonjudgmental, agreeable interactions are *less* productive than sessions where criticism is accepted and even solicited as part of the game.[12] If you or one of your study buddies thinks something is wrong in your understanding, it's important to be able to plainly say so, and to hash out why it's wrong without worrying about hurt feelings. Of course, you don't want to go about gratuitously bashing other people, but too much concern for creating a "safe environment" for criticism actually kills the ability to think constructively and creatively, because you're focusing on the other people rather than the material at hand. Like Feynman, you want to remember that criticism, whether you are giving or receiving it, isn't really about you. It's about what you are trying to understand. In a related vein, people often don't realize that competition can be a good thing—competition is an intense form of collaboration that can help bring out people's best.

Brainstorming buddies, friends, and teammates can help in another way. You often don't mind looking stupid in front of friends. But you don't want to look *too* stupid—at least, not too often. Studying with others, then, can be a little bit like practicing in front of an audience. Research has shown that such public practice makes it easier for you to think on your feet and react well in stressful situations such as those you encounter when you take tests or give a presentation.[13] There is yet another value to study buddies—this relates to when credible sources are in error. Inevitably, no matter how good they are, your instructor—or the book—will make a mistake. Friends can help validate and untangle the resulting confusion and prevent hours of following false leads as you try to find a way to explain something that's flat-out wrong.

But a final word of warning: study groups can be powerfully effective for learning in math, science, engineering, and technology. If study sessions turn into socializing occasions, however, all bets are off. Keep small talk to a minimum, get your group on track, and finish your work.[14] If you find that your group meetings start five to fifteen minutes late, members haven't read the material, and the conversation consistently veers off topic, find yourself another group.

TEAMWORK FOR INTROVERTS

"I'm an introvert and I don't like working with people. But when I wasn't doing so well in my college engineering classes (back in the 1980s), I decided that I needed a second pair of eyes, although I still didn't want to work with anyone. Since we didn't have online chatting back then, we wrote notes on each other's doors in the dorms. My classmate Jeff and I had a system: I would write '1) 1.7 m/s'—meaning that the answer to homework problem one was 1.7 meters per second. Then I'd get back from a shower and see that Jeff had written, 'No, 1) 11 m/s.' I'd desperately go through my own work and find a mistake, but now I had 8.45 m/s. I'd go down to Jeff's room and we'd argue intensively with both our solutions out while he had a guitar slung around his shoulder. Then we'd both go back to our own work on our own time and I'd suddenly see that the answer was 9.37 m/s, and so would he, and we'd both get 100 percent on the homework assignment. As you can see, there are ways to work with others that require only minimal interaction if you don't like working in groups."

—Paul Blowers, University Distinguished Professor (for extraordinary teaching), University of Arizona

---

SUMMING IT UP

- The focused mode can allow you to make critical errors even though you feel confident you've done everything correctly. Rechecking your work can allow you to get a broader perspective on it, using slightly different neural processes that can allow you to catch blunders.

- Working with others who aren't afraid to disagree can:
  - Help you catch errors in your thinking.
  - Make it easier for you to think on your feet and react well in stressful situations.
  - Improve your learning by ensuring that you really understand what you are explaining to others and reinforcing what you know.
  - Build important career connections and help steer you toward better choices.
- Criticism in your studies, whether you are giving or receiving it, shouldn't be taken as being about you. It's about what you are trying to understand.
- It is easiest of all to fool yourself.

> **PAUSE AND RECALL**
>
> Close the book and look away. What were the main ideas of this chapter? Try recalling some of these ideas when you are around friends—it will also help your friends to know how valuable their interactions with you actually are!

ENHANCE YOUR LEARNING

1. Describe an example of how you were absolutely 100 percent certain of something and were later proven wrong. As a result of this and similar incidents, do you think you are more capable now of accepting criticism of your ideas from others?

2. How could you make your study sessions with classmates more effective?

3. How would you handle it if you found yourself in a group that seemed to focus on other issues besides your studies?

INSIGHTS ON LEARNING FROM PHYSICS PROFESSOR BRAD ROTH, A FELLOW OF THE AMERICAN PHYSICAL SOCIETY AND CO-AUTHOR OF *INTERMEDIATE PHYSICS FOR MEDICINE AND BIOLOGY*

Brad Roth and his dog Suki, enjoying the Michigan fall color.

"One thing I stress in my classes is to think before you calculate. I really hate the 'plug and chug' approach that many students use. Also, I find myself constantly reminding students that equations are NOT merely expressions you plug numbers into to get other numbers. Equations tell a story about how the physical world works. For me, the key to understanding an equation in physics is to see the underlying story. A qualitative understanding of an equation is more important than getting quantitatively correct numbers out of it.

"Here are a few more tips:

1. *"Often, it takes way less time to check your work than to solve a problem. It is a pity to spend twenty minutes solving a problem and then get it wrong because you did not spend two minutes checking it.*

2. *"Units of measurement are your friend. If the units don't match on each side of an equation, your equation is not correct. You can't add something with units of seconds to something with units of meters. It's like adding apples and rocks—nothing edible comes of it. You can look back at your work, and if you find the place where the units stop matching, you probably will find your mistake. I have been asked to review research papers that are submitted to professional journals that contain similar unit errors.*

3. *"You need to think about what the equation means, so that your math result and your intuition match. If they don't match,*

then you have either a mistake in your math or a mistake in your intuition. Either way, you win by figuring out why the two don't match.

4. *"(Somewhat more advanced) For a complicated expression, take limiting cases where one variable or another goes to zero or infinity, and see if that helps you understand what the equation is saying."*

....

{ 17 }

test taking

We've mentioned it earlier, but it's worth repeating, in bold letters: **Testing is itself an extraordinarily powerful learning experience.** This means that the effort you put into test taking, including the preliminary mini-tests of your recall and your ability to problem-solve during your preparation, is of fundamental importance. If you compare how much you learn by spending one hour studying versus one hour taking a test on that same material, you will retain and learn far more as a result of the hour you spent taking a test. Testing, it seems, has a wonderful way of concentrating the mind.

Virtually everything we've talked about in this book has been designed to help make the testing process seem straightforward and natural—simply an extension of the normal procedures you use to learn the material. So it's time now to cut directly to one of the central features of this chapter and the entire book—a checklist you can use to see whether your preparation for test taking is on target.

TEST PREPARATION CHECKLIST

Professor Richard Felder is a legend among engineering educators—he has arguably done as much as or more than any educator in this century to help students worldwide to excel in math and science.[1] One of the simplest and perhaps most effective techniques Dr. Felder has used to help students is laid out in a memo he wrote to students who have been disappointed with their test grades.[2]

"Many of you have told your instructor that you understood the course material much better than your last test grade showed, and some of you asked what you should do to keep the same thing from happening on the next test.

"Let me ask you some questions about how you prepared for the test. Answer them as honestly as you can. If you answer 'No' to many of them, your disappointing test grade should not be too surprising. If there are still a lot of 'Nos' after the next test, your disappointing grade on that test should be even less surprising. If your answer to most of these questions is 'Yes' and you still got a poor grade, something else must be going on. It might be a good idea for you to meet with your instructor or a counselor to see if you can figure out what it is.

"You'll notice that several of the questions presume that you're working with classmates on the homework—either comparing solutions you first obtained individually or actually getting together to work out the solutions. Either approach is fine. In fact, if you've been working entirely by yourself and your test grades are unsatisfactory, I would strongly encourage you to find one or two homework and study partners to work with before the next test. (Be careful about the second approach, however; if what you're doing is mainly watching others work out solutions, you're probably doing yourself more harm than good.)

"The answer to the question 'How should I prepare for the test?' becomes clear once you've filled out the checklist. You should:

Do Whatever It Takes to Be Able to Answer "Yes" to Most of the Questions.

Test Preparation Checklist

Answer "Yes" only if you *usually* did the things described (as opposed to occasionally or never).

Homework

__Yes __No 1. Did you make a serious effort to understand the text? (Just hunting for relevant worked-out examples doesn't count.)

__Yes __No 2. Did you work with classmates on homework problems, or at least check your solutions with others?

__Yes __No 3. Did you attempt to outline every homework problem solution before working with classmates?

Test Preparation

The more "Yes" responses you recorded, the better your preparation for the test. If you recorded two or more "No" responses, think seriously about making some changes in how you prepare for the next test.

__Yes __No 4. Did you participate actively in homework group discussions (contributing ideas, asking questions)?

__Yes __ No 5. Did you consult with the instructor or teaching assistants when you were having trouble with something?

__Yes __ No 6. Did you understand ALL of your homework problem solutions when they were handed in?

__Yes __No 7. Did you ask in class for explanations of homework problem solutions that weren't clear to you?

__Yes __No 8. If you had a study guide, did you carefully go through it before the test and convince yourself that you could do everything on it?

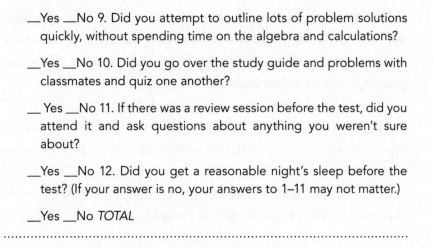

__Yes __No 9. Did you attempt to outline lots of problem solutions quickly, without spending time on the algebra and calculations?

__Yes __No 10. Did you go over the study guide and problems with classmates and quiz one another?

__ Yes __No 11. If there was a review session before the test, did you attend it and ask questions about anything you weren't sure about?

__Yes __No 12. Did you get a reasonable night's sleep before the test? (If your answer is no, your answers to 1–11 may not matter.)

__Yes __No *TOTAL*

The *Hard-Start–Jump-to-Easy* Technique

The classic way students are taught to approach tests in math and science is to tackle the easiest problems first. This is based on the notion that that by the time you've finished the relatively simple problems, you'll be confident in handling the more difficult.

This approach works for some people, mostly because anything works for *some* people. Unfortunately, however, for most people it's counterproductive. Tough problems often need lots of time, meaning you'd want to start on them first thing on a test. Difficult problems also scream for the creative powers of the diffuse mode. But to access the diffuse mode, you need to *not* be focusing on what you want so badly to solve!

What to do? Easy problems first? Or hard?

The answer is to start with the hard problems—but quickly jump to the easy ones. Here's what I mean.

When the test is handed out to you, first take a quick look to get a sense of what it involves. (You should do this in any case.) Keep your eye out for what appears to be the hardest problem.

Then **when you start working problems, start first with what appears to be the hardest one. But steel yourself to pull away within the first minute or two if you get stuck or get a sense that you might not be on the right track.**

This does something exceptionally helpful. "Starting hard" loads the first, most difficult problem in mind, and then switches attention away from it. *Both these activities can help allow the diffuse mode to begin its work.*

If your initial work on the first hard problem has unsettled you, turn next to an easy problem, and complete or do as much as you can. Then move next to another difficult-looking problem and try to make a bit of progress. Again, change to something easier as soon as you feel yourself getting bogged down or stuck.

....

"With my students, I talk about good worry and bad worry. Good worry helps provide motivation and focus while bad worry simply wastes energy."

—Bob Bradshaw, Professor of Math, Ohlone College

....

When you return to the more difficult problems, you'll often be pleased that the next step or steps in the problem will seem more obvious to you. You may not be able to get all the way to the end immediately, but at least you can get further before you switch to something else on which you can make progress.

In some sense, with this approach to test taking, you're being like an efficient chef. While you're waiting for a steak to fry, you can swiftly slice the tomato garnish, then turn to season the soup, and then stir the sizzling onions. The *hard-start–jump-to-easy* technique may make more efficient use of your brain by allowing different parts of the brain to work simultaneously on different thoughts.[3]

Using the hard-start–jump-to-easy technique on tests guarantees you will have at least a little work done on every problem. It is also a valuable technique for helping you avoid *Einstellung*—getting stuck in the wrong approach—because you have a chance to look at the problems from differing perspectives at different times. All this is particularly important if your instructor gives you partial credit.

The only challenge with this approach is that *you must have the self-discipline to pull yourself off a problem once you find yourself stuck for a minute or two.* For most students, it's easy. For others, it takes discipline and willpower. In any case, by now you are very aware that misplaced persistence can create unnecessary challenges with math and science.

This may be why test takers sometimes find that the solution pops to mind right as they walk out the door. When they gave up, their attention switched, allowing the diffuse mode the tiny bit of traction it needed to go to work and return the solution. Too late, of course.

Sometimes people are concerned that starting a problem and then pulling away from it might cause confusion in an examination. This doesn't seem to be a problem for most people; after all, chefs learn to bring various facets of a dinner together. But if you still have worries about whether this strategy might work for you, try it first on homework problems.

Be aware of some occasions when hard-start–jump-to-easy might not be appropriate. If the instructor gives only a few points for a really difficult problem (some instructors like to do this), you may wish to concentrate your efforts elsewhere. Some computerized licensure examinations don't allow for backtracking, so your best bet when facing a tough question is simply to take a deep breath or two from the belly (make sure to breathe out all the way, also) and do your best. And if you haven't prepared well for the test, then all bets are off. Take what simple points you can.

DEALING WITH PANIC BEFORE A TEST

"I tell my students to *face your fears*. Often, your worst fear is not getting the grades you need for your chosen career. How can you handle this? Simple. Have a plan B for an alternative career. Once you have a plan for the worst contingency, you'll be surprised to see that the fear will begin to subside.

"Study hard up until the day of the test, and then let it go. Tell yourself, 'Oh, well, let me just see how many questions I can get right. I can always pursue my other career choice.' That helps release stress so you actually do better and get closer to your first career choice."

—*Tracey Magrann, Professor of Biological Sciences, Saddleback College*

Why Anxiety Can Arise on Tests and How to Deal with It

If you're a stressed-out test taker, keep in mind that the body puts out chemicals, such as cortisol, when it is under stress. This can cause sweaty palms, a racing heart, and a knot in the pit of your stomach. But interestingly, research finds that it's how you *interpret* those symptoms—the story you tell yourself about why you are stressed—that makes all the difference. If you shift your thinking from "this test has made me afraid" to "this test has got me excited to do my best!" it can make a significant improvement in your performance.[4]

Another good tip for panicky test takers is to momentarily turn your attention to your breathing. Relax your stomach, place your hand on it, and slowly draw a deep breath. Your hand should move out, even as your whole chest is moving outward like an expanding barrel.

By doing this type of deep breathing, you are sending oxygen to critical areas of your brain. This signals that all is well and helps calm you down. But don't just start this breathing on the day of the test. If you have practiced this breathing technique in the weeks before—just a minute or two here and there is all it takes—you will slide more easily into the breathing pattern during the test. (Remember, practice makes permanent!) It's particularly helpful to move into the deeper breathing pattern in those final anxious moments before a test is handed out. (And yes, if you're interested, there are dozens of apps to help you.)

Another technique involves mindfulness.[5] In this technique, you learn to distinguish between a naturally arising thought (I have a big test next week) and an emotional projection that can tag along after that initial thought (If I flunk the test, I will wash out of the program, and I'm not sure what I'll do then!). These tagalong thoughts, it seems, are projections that arise as glimmers from the diffuse mode. Even a few weeks of simple practice in learning to reframe these thoughts and feelings as simple mental tagalong projections seems to help ease and quiet the mind. Reframing your reaction to such intrusive thoughts works much better than simply trying to suppress them. Students who spent a few weeks practicing with the mindfulness approach performed better on their tests, experiencing fewer distracting thoughts.

Now you can see why waiting until the end of the test to work on the hardest questions can lead to problems. Just when you are increasingly stressed out because you are running out of time, you are also suddenly facing the toughest problems! As your stress levels soar, you concentrate intently, thinking that focused attention will solve your problems, but of course, your focus instead prevents the diffuse mode from being able to go to work.

The result? "Paralysis by analysis."[6] The "hard-start–jump-to-easy" technique helps prevent this.

MULTIPLE "GUESS" AND PRACTICE TESTS . . . A FEW TIPS

"When I give multiple-choice tests, I sometimes find that students fail to fully grasp what the question is asking before they barge ahead, reading the answer options. I advise them to cover up the answers and to try to recall the information so they can answer the question on their own first.

"When my students complain that the practice test was *waaaaay* easier than the real one, I ask: What are the confounding variables that make the two situations different? When you took the practice test, were you at home relaxing with tunes on? Taking it with a fellow student? No time limit? Answer key and class materials at hand? These circumstances are not exactly like a crowded classroom with a clock ticking away. I actually encourage those with test anxiety to bring their practice test to another class (big classes where one can slip right in and sit at the back unnoticed) and try taking it there."

—*Susan Sajna Hebert, Professor of Psychology,*
Lakehead University

Final Thoughts on Testing

The day before a test (or tests), have a quick look over the materials to brush up on them. You'll need both your focused-mode and diffuse mode "muscles" the next day, so you don't want to push your brain too hard. (You wouldn't run a ten-mile race the day before running a marathon.) Don't feel guilty if you can't seem to get yourself to work too hard the day before a big examination. If you've prepared properly, this is a natural reaction: You are subconsciously pulling back to conserve mental energy.

While taking a test, you should also remember how your mind can trick you into thinking what you've done is correct, even if it isn't. This means that, **whenever possible, you should blink, shift your attention, and then** *double-check your answers* **using a big-picture perspective, asking yourself, "Does this really make** *sense*?" There is often more than one way to solve a problem, and checking your answers from a different perspective provides a golden opportunity for verifying what you've done.

If there's no other way to check except to step back through your logic, keep in mind that simple issues like missed minus signs, incorrectly added numbers, and "dropped atoms" have tripped up even the most advanced mathematics, science, and engineering students. Just do your best to catch them. In science classes, having your units of measurement match on each side of the equation can provide an important clue about whether you've done the problem correctly.

The order in which you work tests is also important. Students generally work tests from front to back. When you are checking your work, if you start more toward the back and work toward the front, it sometimes seems to give your brain a fresher perspective that can allow you to more easily catch errors.

Nothing is ever certain. Occasionally you can study hard and the test gods simply don't cooperate. But if you prepare well by practicing and by building a strong mental library of problem-solving techniques, and approach test taking wisely, you will find that luck will increasingly be on your side.

SUMMING IT UP

- Not getting enough sleep the night before a test can negate any other preparation you've done.
- Taking a test is serious business. Just as fighter pilots and doctors go through checklists, going through your own test preparation checklist can vastly improve your chances of success.
- Counterintuitive strategies such as the hard-start–jump-to-easy technique can give your brain a chance to reflect on harder challenges even as you're focusing on other, more straightforward problems.
- The body puts out chemicals when it is under stress. How you interpret your body's reaction to these chemicals makes all the difference. If you shift your thinking from "This test has made me afraid" to "This test has got me excited to do my best!" it helps improve your performance.
- If you are panicked on a test, momentarily turn your attention to your breathing. Relax your stomach, place your hand on it, and slowly draw a deep breath. Your hand should move outward, and your whole chest should expand like a barrel.
- Your mind can trick you into thinking that what you've done is correct, even if it isn't. This means that, whenever possible, you should blink, shift your attention, and then double-check your answers using a big-picture perspective, asking yourself, "Does this really make *sense*?"

{ **PAUSE AND RECALL**

Close the book and look away. What were the main ideas of this chapter? What new ideas will be particularly important for you to try related to testing?

ENHANCE YOUR LEARNING

1. What is the one extraordinarily important preparation step for taking a test? (Hint: If you don't take this step, nothing else you do to prepare for the test matters.)

2. Explain how you would determine whether it is time to pull yourself off a difficult problem on a test when you are using the hard-start–jump-to-easy technique.

3. A deep-breathing technique was suggested to help with feelings of panic. Why do you think the discussion emphasized breathing so that the belly rises, rather than just the upper chest?

4. Why would you want to try to shift your attention momentarily before rechecking your answers on a test?

PSYCHOLOGIST SIAN BEILOCK ON HOW TO PREVENT
THE DREADED "CHOKE"

Sian Beilock is a psychology professor at the University of Chicago. She is one of the world's leading experts on how to reduce feelings of panic under high-stakes conditions, and is the author of the book *Choke: What the Secrets of the Brain Reveal about Getting It Right When You Have To.*[7]

"High-stakes learning and performance situations can put you under a lot of stress. However, there is a growing body of research showing that fairly simple psychological interventions can lower your anxiety about tests and boost what you learn in the classroom. These interventions don't teach academic content; they target your attitudes.

"My research team has found that if you write about your thoughts and feelings about an upcoming test immediately before you take the test, it can lessen the negative impact of pressure on performance. We think that writing helps to release negative thoughts from mind, making them less likely to pop up and distract you in the heat of the moment.

"The minor stress of many self-tests as you master the material can also prepare you for the more intense stress of real tests. As you've learned in this book, testing yourself while you are learning is a great way to commit information to mind, making it easier to fish out in the heat of a high-stakes exam.

"It's also true that negative self-talk—that is, negative thoughts arising from your own mind—can really hurt your performance, so make sure that what you say and think about yourself as you are preparing for tests is always upbeat. Cut yourself off in midthought if need be to prevent negativity, even if you feel the dragons of doom await you. If you flub a problem, or even many problems, keep your spirits up and turn your focus to the next problem.

"Finally, one reason students sometimes choke on a test is that they

frantically dive right in to solving a problem before they've really thought about what they are facing. Learning to pause for a few seconds before you start solving a problem or when you hit a roadblock can help you see the best solution path—this can help prevent the ultimate choking feeling when you suddenly realize you've spent a lot of time pursuing a dead end.

"You can definitely learn to keep your stress within bounds. Surprisingly, you wouldn't want to eliminate stress altogether, because a little stress can help you perform at your best when it matters most.

"Good luck!"

{ 18 }

unlock your potential

Richard Feynman, the bongo-playing, Nobel Prize–winning physicist, was a happy-go-lucky guy. But there were a few years—the
best and worst of his life—when his exuberance was challenged.

In the early 1940s, Feynman's beloved wife, Arlene, lay in a distant hospital, deathly ill with tuberculosis. He only rarely could get
away to see her because he was in the isolated New Mexico town
of Los Alamos, working on one of the most important projects of
World War II—the top-secret Manhattan Project. Back then, Feynman was nobody famous. No special privileges were afforded him.

To help keep his mind occupied when his workday ended and
anxiety or boredom reared its head, Feynman began a focused effort to peer into people's deepest, darkest secrets: He began figuring out how to open safes.

Becoming an accomplished safecracker isn't easy. Feynman
developed his intuition, mastering the internal structures of the
locks, practicing like a concert pianist so his fingers could swiftly

run through remaining permutations if he could discover the first numbers of a combination.

Eventually, Feynman happened to learn of a professional locksmith who had recently been hired at Los Alamos—a real expert who could open a safe in seconds.

An expert, right at hand! Feynman realized if only he could befriend this man, the deepest secrets of safecracking would be his.

IN THIS BOOK we've explored new ways of looking at how you learn. Sometimes, as we've discovered, **your desire to figure things out** *right now* **is what** *prevents* **you from being able to figure things out.** It's almost as if, when you reach too quickly with your right hand, your left hand automatically latches on and holds you back.

Great artists, scientists, engineers, and chess masters like Magnus Carlsen tap into the natural rhythm of their brains by first intently focusing their attention, working hard to get the problem well in mind. Then they switch their attention elsewhere. This alternation between focused and diffuse methods of thinking allows thought clouds to drift more easily into new areas of the brain. Eventually, snippets of these clouds—refined, refluffed—can return with useful parts of a solution.

Reshaping your brain is under your control. The key is patient persistence—working knowledgeably *with* **your brain's strengths and weaknesses.**

You can improve your focusing ability by gently redirecting your responses to interrupting cues like your phone's ring or the beep of a text message. The Pomodoro—a brief, timed period of focused attention—is a powerful tool in diverting the well-meaning zombies of your habitual responses. Once you've done a bout of hard, focused work, you can then really savor the mental relaxation that follows.

The result of weeks and months of gradual effort? Sturdy neural structures with well-cured mortar laid between each new learning period. Learning in this way, with regular periods of relaxation between times of focused attention, not only allows us to have more fun, but also allows us to learn more deeply. The relaxation periods provide time to gain perspective—to synthesize both the context and the big picture of what we are doing.

Be mindful that parts of our brain are wired to believe that whatever we've done, no matter how glaringly wrong it might be, is just fine, *thank you very much.* Indeed, our ability to fool ourselves is part of why we check back—*Does this really make sense?*—before turning in an examination. By ensuring that we step back and take fresh perspectives on our work, by testing ourselves through recall, and by allowing our friends to question us, we can better catch our illusions of competence in learning. It is these illusions, as much as any real lack of understanding, that can trip us up en route to success in studying math and science.

Rote memorization, often at the last minute, has given many lower-level learners the illusory sense that they understand math and science. As they climb to higher levels, their weak understanding eventually crumbles. But our growing understanding of how the mind truly learns is helping us move past the simplistic idea that memorization is always bad. We now know that deep, practiced internalization of well-understood chunks is *essential* to mastering math and science. We also know that, just as athletes can't properly develop their muscles if they train in last-minute cramming sessions, students in math and science can't develop solid neural chunks if they procrastinate in their studies.

No matter what our age and degree of sophistication, parts of our brain remain childlike. This means that we sometimes can become frustrated, a signal to us to take a breather. But our ever-present inner child also gives us the potential to let go and use our creativity to help us visualize, remember, make friends with, and truly understand concepts in math and science that at first can seem terribly difficult.

In a similar way, we've found that persistence can sometimes be misplaced—that relentless focus on a problem blocks our ability to solve that problem. At the same time, big-picture, long-term persistence is key to success in virtually any domain. This kind of long-term stick-to-it-iveness is what can help get us past the inevitable naysayers or unfortunate vicissitudes of life that can temporarily make our goals and dreams seem too far to grasp.

A central theme of this book is the paradoxical nature of learning. Focused attention is indispensable for problem solving—yet it can also block our ability to solve problems. Persistence is key—but it can also leave us unnecessarily pounding our heads. Memorization is a critical aspect of acquiring expertise—but it can also keep us focused on the trees instead of the forest. Metaphor allows us to acquire new concepts—but it can also keep us wedded to faulty conceptions.

Study in groups or alone, start hard or start easy, learn concretely or in abstract, success or failure . . . In the end, integrating the many paradoxes of learning adds value and meaning to everything we do.

Part of the magic long used by the world's best thinkers has been to simplify—to put things into terms that even a kid brother or sister can understand. This, indeed, was Richard Feynman's approach; he challenged some of the most esoteric theoretical mathematicians he knew to put their complicated theories in simple terms.

It turned out they could. You can, too. And like both Feynman and Santiago Ramón y Cajal, you can use the strengths of learning to help reach your dreams.

· · ·

AS FEYNMAN CONTINUED to refine his safecracking skills, he be-friended the professional locksmith. Through time and talk, Feyn-man gradually swept away superficial pleasantries, digging deeper and deeper so that he could understand the nuance behind what he saw to be the locksmith's utter mastery.

Late one night, at long last, that most valuable of arcane knowl-edge became clear.

The locksmith's secret was that he was privy to the manufacturers' de-fault settings.

By knowing the default settings, the locksmith was often able to slip into safes that had been left unchanged since they'd arrived from the manufacturer. Whereas everyone thought that safecrack-ing wizardry was involved, it was a simple understanding of how the device arrived from the manufacturer that was fundamental.

Like Feynman, you can achieve startling insights into how to understand more simply, easily, and with less frustration. By under-standing your brain's default settings—the natural way it learns and thinks—and taking advantage of this knowledge, you, too, can be-come an expert.

In the beginning of the book, I mentioned that there are simple mental tricks that can bring math and science into focus, tricks that are helpful not only for people who are bad at math and science but also for those who already good at it. You've walked through all these tricks in the course of reading this book. But, as you now know, nothing beats grasping the chunked and simplified essence. So what follows sums up my final thoughts—the chunked essence of some of the central ideas in this book, distilled into the ten best and worst rules of studying.

Remember—Lady Luck favors the one who tries. A little insight into learning how to learn best doesn't hurt, either.

TEN RULES OF GOOD STUDYING

1. **Use recall.** After you read a page, look away and recall the main ideas. Highlight very little, and never highlight anything you haven't put in your mind first by recalling. Try recalling main ideas when you are walking to class or in a different room from where you originally learned it. An ability to recall—to generate the ideas from inside yourself—is one of the key indicators of good learning.

2. **Test yourself.** On everything. All the time. Flash cards are your friend.

3. **Chunk your problems.** Chunking is understanding and practicing with a problem solution so that it can all come to mind in a flash. After you solve a problem, rehearse it. Make sure you can solve it cold—every step. Pretend it's a song and learn to play it over and over again in your mind, so the information combines into one smooth chunk you can pull up whenever you want.

4. **Space your repetition.** Spread out your learning in any subject a little every day, just like an athlete. Your brain is like a muscle—it can handle only a limited amount of exercise on one subject at a time.

5. **Alternate different problem-solving techniques during your practice.** Never practice too long at any one session using only one problem-solving technique—after a while, you are just mimicking what you did on the previous problem. Mix it up and work on different types of problems. This teaches you both *how* and *when* to use a technique. (Books generally are not set up this way, so you'll need to do this on your own.) After every assignment and test, go over your errors, make sure you understand why you made them, and then rework your solutions. To study most effectively, handwrite (don't type) a problem on one side of a flash card and the solution on the other. (Handwriting builds stronger neural structures in memory than typing.) You might also photo-

graph the card if you want to load it into a study app on your smartphone. Quiz yourself randomly on different types of problems. Another way to do this is to randomly flip through your book, pick out a problem, and see whether you can solve it cold.

6. **Take breaks.** It is common to be unable to solve problems or figure out concepts in math or science the first time you encounter them. This is why a little study every day is much better than a lot of studying all at once. When you get frustrated with a math or science problem, take a break so that another part of your mind can take over and work in the background.

7. **Use explanatory questioning and simple analogies.** Whenever you are struggling with a concept, think to yourself, *How can I explain this so that a ten-year-old could understand it?* Using an analogy really helps, like saying that the flow of electricity is like the flow of water. Don't just think your explanation—say it out loud or put it in writing. The additional effort of speaking and writing allows you to more deeply encode (that is, convert into neural memory structures) what you are learning.

8. **Focus.** Turn off all interrupting beeps and alarms on your phone and computer, and then turn on a timer for twenty-five minutes. Focus intently for those twenty-five minutes and try to work as diligently as you can. After the timer goes off, give yourself a small, fun reward. A few of these sessions in a day can really move your studies forward. Try to set up times and places where studying—not glancing at your computer or phone—is just something you naturally do.

9. **Eat your frogs first.** Do the hardest thing earliest in the day, when you are fresh.

10. **Make a mental contrast.** Imagine where you've come from and contrast that with the dream of where your studies will take you. Post a picture or words in your workspace to remind you of your dream. Look at that when you find your motivation lagging. This work will pay off both for you and those you love!

TEN RULES OF BAD STUDYING

Avoid these techniques—they can waste your time even while they fool you into thinking you're learning!

1. **Passive rereading**—sitting passively and running your eyes back over a page. Unless you can *prove* that the material is moving into your brain by recalling the main ideas without looking at the page, rereading is a waste of time.

2. **Letting highlights overwhelm you.** Highlighting your text can fool your mind into thinking you are putting something in your brain, when all you're really doing is moving your hand. A little highlighting here and there is okay—sometimes it can be helpful in flagging important points. But if you are using highlighting as a memory tool, make sure that what you mark is also going into your brain.

3. **Merely glancing at a problem's solution and thinking you know how to do it.** This is one of the worst errors students make while studying. You need to be able to *solve* a problem step-by-step, without looking at the solution.

4. **Waiting until the last minute to study.** Would you cram at the last minute if you were practicing for a track meet? Your brain is like a muscle—it can handle only a limited amount of exercise on one subject at a time.

5. **Repeatedly solving problems of the same type that you already know how to solve.** If you just sit around solving similar problems during your practice, you're not actually preparing for a test—it's like preparing for a big basketball game by just practicing your dribbling.

6. **Letting study sessions with friends turn into chat sessions.** Checking your problem solving with friends, and quizzing one another on what you know, can make learning more enjoyable, expose flaws in your thinking, and deepen your learning. But if your joint study sessions turn to fun before the work is done, you're wasting your time and should find another study group.

7. **Neglecting to read the textbook before you start working problems.** Would you dive into a pool before you knew how to swim? The textbook is your swimming instructor—it guides you toward the answers. You will flounder and waste your time if you don't bother to read it. Before you begin to read, however, take a quick glance over the chapter or section to get a sense of what it's about.

8. **Not checking with your instructors or classmates to clear up points of confusion.** Professors are used to lost students coming in for guidance—it's our job to help you. The students we worry about are the ones who don't come in. Don't be one of those students.

9. **Thinking you can learn deeply when you are being constantly distracted.** Every tiny pull toward an instant message or conversation means you have less brain power to devote to learning. Every tug of interrupted attention pulls out tiny neural roots before they can grow.

10. **Not getting enough sleep.** Your brain pieces together problem-solving techniques when you sleep, and it also practices and repeats whatever you put in mind before you go to sleep. Prolonged fatigue allows toxins to build up in the brain that disrupt the neural connections you need to think quickly and well. If you don't get a good sleep before a test, NOTHING ELSE YOU HAVE DONE WILL MATTER.

PAUSE AND RECALL

Close the book and look away. What were the most important ideas in this book? As you reflect, consider also how you will use these ideas to help reshape your learning.

afterword

My eighth-grade math and science teacher had a powerful impact on my life. He plucked me from the back of the class and motivated me to strive for excellence. I repaid him in high school by getting a D in geometry—twice. I just couldn't get the material on my own, and I didn't have the luxury of a great teacher to prod me in the ways I needed. Eventually, in college, I figured it out. But it was a frustrating journey. I wish I'd had a book like this back then.

Flash forward a decade and a half. My daughter turned math homework into a form of torture Dante would be too shy to write about. She would hit a wall and then hit it again and again. When she finally finished crying, she would circle around and eventually figure it out. But I could never get her to just back off and regroup without the drama. I let her read this book. The first thing she said was, "I wish I'd had this book when I was in school!"

There has long been a stream of potentially productive study advice coming from scientists. Unfortunately, it has seldom been translated so the average student can easily grasp and use it. Not every scientist has a knack for translation, and not every writer has a firm grasp of the science. In this book, Barbara Oakley threaded this needle beautifully. Her use of vivid examples and explanations of the strategies reveals not only how useful but how credible these ideas are. When I asked my daughter why she liked the advice in the book, even though I had mentioned several of the techniques to her when she was in middle school, she said, "She tells you why and it makes sense." Another hit to my parental ego!

Now that you have read this book, you have been exposed to some simple yet potentially powerful strategies—strategies, by the way, that could benefit you in more than just math and science. As you've discovered, these strategies grew from considerable evidence about how the human mind works. The interplay between emotion and cognition, though seldom put into words, is an essential component to all learning. In her own way, my daughter pointed out that studying isn't just about the strategies. You have to be convinced that those strategies can actually work. The clear and compelling evidence you read in this book should give you the confidence to try techniques without the doubt and resistance that often sabotages our best efforts. Learning is, of course, personally empirical. The ultimate evidence will come when you evaluate your performance and attitude once you earnestly deploy these strategies.

I am now a college professor and I have advised thousands of students over the years. Many students try to avoid math and science because they "are not good at it" or "don't like it." My advice to these students has always been the same advice I gave my daughter: "Get good at it, and then see if you still want to quit." After all, isn't education supposed to be about getting good at challenging things?

Remember how difficult learning to drive was? Now, it is almost automatic and gives you a sense of independence you will value throughout their adult life. By being open to new strategies like the ones in this book, learners now have the opportunity to move past anxiety and avoidance toward mastery and confidence.

It is now up to you: Get good!

—*David B. Daniel, Ph.D.*
Professor, Psychology Department
James Madison University

acknowledgments

In acknowledging the support of these individuals, I would like to make clear that any errors of fact or interpretation in this book are my own. To anyone whose name I might have inadvertently omitted, my apologies.

Underlying this entire effort have been the unwavering support, encouragement, enthusiasm, and superb insight of my husband, Philip Oakley. We met thirty years ago at the South Pole Station in Antarctica—truly I had to go to the ends of the earth to meet that extraordinary man. He is my soul mate and my hero. (And, in case you might have wondered, he is also the man in the puzzle.)

A master mentor throughout my teaching career is Dr. Richard Felder—he has made an enormous difference in how that career unfolded. Kevin Mendez, this book's artist, has done an incredible job in rendering the illustrations—I am in awe of his artistic ability and vision. Our elder daughter, Rosie Oakley, has provided keen insight and unbelievable encouragement throughout the development of this book. Our younger daughter, Rachel Oakley, has always been a pillar of support in our lives.

My good friend Amy Alkon has what amounts to editorial X-ray vision—she has an uncanny ability to ferret out areas for improvement, and with her help this book has reached a far higher level of clarity, accuracy, and wit. My old friend Guruprasad Madhavan of the National Academy of Sciences has helped me see the big-picture implications, as has our mutual friend Josh Brandoff. Writing coach Daphne Gray-Grant has also been a great supporter in the development of this work.

I would especially like to acknowledge the foundational efforts of Rita Rosenkranz, a literary agent of unparalleled excellence. At Penguin, my deepest thanks and appreciation go to Sara Carder and Joanna Ng, whose vision, editorial acumen, and vast expertise with publishing have helped immeasurably in strengthening this book. In particular, I can only wish that every author would be so lucky as to work with someone who possesses Joanna Ng's extraordinary editorial talent. I would also like to extend my thanks to Amy J. Schneider, whose copyediting abilities have been a wonderful boon for this work.

Special thanks go to Paul Kruchko, whose simple question about how I changed got me started on this book. Dante Rance at the Interlibrary Loan Department has continually gone well above and beyond the call of duty; my thanks as well to the supremely capable Pat Clark. Many colleagues have been very supportive in this work, particularly Professors Anna Spagnuolo, László Lipták, and Laura Wicklund in math; Barb Penprase and Kelly Berishaj in nursing; Chris Kobus, Mike Polis, Mohammad-Reza Siadat, and Lorenzo Smith in engineering; and Brad Roth in physics. Aaron Bird, U.S. training manager for CD-adapco, and his colleague Nick Appleyard, vice president at CD-adapco, have both been of exceptional help. I would also like to thank Tony Prohaska for his keen editorial eye.

The following people have also been remarkably helpful in

sharing their expertise: Sian Beilock, Marco Bellini, Robert M. Bilder, Maria Angeles Ramón y Cajal, Norman D. Cook, Terrence Deacon, Javier DeFelipe, Leonard DeGraaf, John Emsley, Norman Fortenberry, David C. Geary, Kary Mullis, Nancy Cosgrove Mullis, Robert J. Richards, Doug Rohrer, Sheryl Sorby, Neel Sundaresan, and Nicholas Wade.

Some of the world's top-ranked university and college professors, as noted on RateMyProfessors.com, have lent invaluable support to this effort. Their expertise includes mathematics, physics, chemistry, biology, science, engineering, business, economics, finance, education, psychology, sociology, nursing, and English. High school teachers from top magnet schools have contributed as well. I would like to particularly acknowledge the assistance of the following individuals, who have read all or portions of the book and provided helpful feedback and insights: Lola Jean Aagaard-Boram, Shaheem Abrahams, John Q. Adams, Judi Addelston, April Lacsina Akeo, Ravel F. Ammerman, Rhonda Amsel, J. Scott Armstrong, Charles Bamforth, David E. Barrett, John Bartelt, Celso Batalha, Joyce Miller Bean, John Bell, Paul Berger, Sydney Bergman, Roberta L. Biby, Paul Blowers, Aby A. Boumarate, Daniel Boylan, Bob Bradshaw, David S. Bright, Ken Broun Jr., Mark E. Byrne, Lisa K. Davids, Thomas Day, Andrew DeBenedictis, Jason Dechant, Roxann DeLaet, Debra Gassner Dragone, Kelly Duffy, Alison Dunwoody, Ralph M. Feather Jr., A. Vennie Filippas, John Frye, Costa Gerousis, Richard A. Giaquinto, Michael Golde, Franklin F. Gorospe IV, Bruce Gurnick, Catherine Handschuh, Mike Harrington, Barrett Hazeltine, Susan Sajna Hebert, Linda Henderson, Mary M. Jensen, John Jones, Arnold Kondo, Patrycja Krakowiak, Anuska Larkin, Kenneth R. Leopold, Fok-Shuen Leung, Mark Levy, Karsten Look, Kenneth MacKenzie, Tracey Magrann, Barry Margulies, Robert Mayes, Nelson Maylone, Melissa McNulty, Elizabeth McPartlan, Heta-Maria

Miller, Angelo B. Mingarelli, Norma Minter, Sherese Mitchell, Dina Miyoshi, Geraldine Moore, Charles Mullins, Richard Musgrave, Richard Nadel, Forrest Newman, Kathleen Nolta, Pierre-Philippe Ouimet, Delgel Pabalan, Susan Mary Paige, Jeff Parent, Vera Pavri, Larry Perez, William Pietro, Debra Poole, Mark Porter, Jeffrey Prentis, Adelaida Quesada, Robert Riordan, Linda Rogers, Janna Rosales, Mike Rosenthal, Joseph F. Santacroce, Oraldo "Buddy" Saucedo, Donald Sharpe, Dr. D. A. Smith, Robert Snyder, Roger Solano, Frances R. Spielhagen, Hilary Sproule, William Sproule, Scott Paul Stevens, Akello Stone, James Stroud, Fabian Hadipriono Tan, Cyril Thong, B. Lee Tuttle, Vin Urbanowski, Lynn Vazquez, Charles Weidman, Frank Werner, Dave Whittlesey, Nader Zamani, Bill Zettler, and Ming Zhang.

The following students have contributed insightful quotes, sidebars, or suggestions for which I am very grateful: Natalee Baetens, Rhiannon Bailey, Lindsay Barber, Charlene Brisson, Randall Broadwell, Mary Cha, Kyle Chambers, Zachary Charter, Joel Cole, Bradley Cooper, Christopher Cooper, Aukury Cowart, Joseph Coyne, Michael Culver, Andrew Davenport, Katelind Davidson, Brandon Davis, Alexander Debusschere, Hannah DeVilbiss, Brenna Donovan, Shelby Drapinski, Trevor Drozd, Daniel Evola, Katherine Folk, Aaron Garofalo, Michael Gashaj, Emanuel Gjoni, Cassandra Gordon, Yusra Hasan, Erik Heirman, Thomas Herzog, Jessica Hill, Dylan Idzkowski, Weston Jeshurun, Emily Johns, Christopher Karras, Allison Kitchen, Bryan Klopp, William Koehle, Chelsey Kubacki, Nikolas Langley-Rogers, Xuejing Li, Christoper Loewe, Jonathon McCormick, Jake McNamara, Paula Meerschaert, Mateusz Miegoc, Kevin Moessner, Harry Mooradian, Nadia Noui-Mehidi, Michael Orrell, Michael Pariseau, Levi Parkinson, Rachael Polaczek, Michelle Radcliffe, Sunny Rishi, Jennifer Rose, Brian Schroll, Paul Schwalbe, Anthony Sciuto, Zac Shaw, David Smith, Kimberlee Somerville, Davy Sproule, P. J. Sproule, Dario Strazimiri, Jonathan

Strong, Jonathan Sulek, Ravi Tadi, Aaron Teachout, Gregory Terry, Amber Trombetta, Rajiv Varma, Bingxu Wang, Fangfei Wang, Jessica Warholak, Shaun Wassell, Malcolm Whitehouse, Michael Whitney, David Wilson, Amanda Wolf, Anya Young, Hui Zhang, and Cory Zink.

endnotes

Chapter 1: Open the Door

1 I'd like to point educators toward the book *Redirect*, by psychology professor Timothy Wilson, which describes the seminal importance of failure-to-success stories (Wilson 2011). Helping students change their inner narratives forms one of the important goals of this book. A leader in describing the importance of change and growth in mindset is Carol Dweck (Dweck 2006).

2 Sklar et al. 2012; Root-Bernstein and Root-Bernstein 1999, chap. 1.

Chapter 2: Easy Does It: Why Trying Too Hard Can Sometimes Be Part of the Problem

1 Default-mode network discussions: Andrews-Hanna 2012; Raichle and Snyder 2007; Takeuchi et al. 2011. More general discussion of resting states: Moussa et al. 2012. In a very different line of investigation, Bruce Mangan has noted that William James's description of the fringe includes the following feature: "There is an 'alternation' of consciousness, such that the fringe briefly but frequently comes to the fore and is dominant over the nucleus of awareness" (Cook 2002, p. 237; Mangan 1993).

2 Immordino-Yang et al. 2012.

3 Edward de Bono is the grand master of creativity studies, and his *vertical* and *lateral* terminology is roughly analogous to my use of the terms *focused* and *diffuse* (de Bono 1970).

  Astute readers will notice my mention that the diffuse mode seems to sometimes work in the background while the focused mode is active. However, research findings show that the default-mode network for example (which is just one of the many resting state networks), seems to go quiet when the focused mode is active. So which is it? My sense as an educator and a learner myself is that some nonfocused activities can continue in the background when focused work is taking place, as long as the focused attention is shifted away from the area of interest. In some sense, then, my use of the term *diffuse mode* might be thought of as "nonfocused mode activities directed toward learning" rather than simply "default-mode network."

4　There are also a few tight links to more distant nodes of the brain, as we'll explore later with the attentional octopus analogy.

5　The diffuse mode may also involve prefrontal areas, but it probably has more connections overall and less filtering out of seemingly irrelevant connections.

6　Psychologist Norman Cook has proposed that "the first elements in a central dogma for human psychology can be expressed as (1) the flow of information between the right and left hemispheres and (2) between the "dominant" [left hemisphere] and the peripheral effector mechanisms used for verbal communication" (Cook 1989, p. 15). But it should also be noted that hemispheric differences have been used to launch countless spurious overextrapolations and inane conclusions (Efron 1990).

7　According to the National Survey of Student Engagement (2012), engineering students spend the most time studying—senior engineering students spend eighteen hours on average per week preparing for class, while senior education students spend fifteen hours and senior social science and business students spend about fourteen hours. In a *New York Times* article titled "Why Science Majors Change Their Minds (It's Just So Darn Hard)," emeritus engineering professor David E. Goldberg has noted that the heavy demands of calculus, physics, and chemistry can initiate the "math-science death march" as students wash out (Drew 2011).

8　For a discussion of evolutionary considerations in mathematical thinking, see Geary 2005, chap. 6.

Of course, many abstract terms aren't related to mathematics. A surprising number of these types of abstract ideas, however, relate to emotions. We may not be able to *see* those terms, but we can *feel* them, or at least important aspects of them.

Terrence Deacon, author of *The Symbolic Species*, notes the inherent complexity of the encryption/decryption problem of mathematics:

"Imagine back when you were first encountering a novel kind of mathematical, like recursive subtraction (i.e., division). Most often this abstract concept is taught by simply having children learn a set of rules for manipulating characters for numbers and operations, then using these rules again and again with different numbers in hopes that this will help them 'see' how this parallels certain physical relationships. We often describe this as initially learning to do the manipulations 'by rote' (which is in my terms indexical learning) and then when this can be done almost mindlessly, we hope that they will see how this corresponds to a physical world process. At some point, if all goes well, kids 'get' the general abstract commonality that lies 'behind' these many individual symbol-to-symbol and formula-to-formula operations. They thus reorganize what they already know by rote according to a higher-order mnemonic that is about these combinatorial possibilities and their abstract correspondence to thing manipulation. This abstraction step is often quite difficult for many kids. But now consider that this same transformation at a yet higher level of abstraction is required to understand calculus. Differentiation is effectively recursive division, and integration is effectively recursive multiplication, each carried out indefinitely, i.e., to infinitesimal values (which is possible because they depend on convergent series, which themselves are only known by inference, not direct inspection). This ability to project what an operation entails when carried out infinitely is what solves Zeno's paradox, which seems impossible when stated in words. But in addition to this difficulty, the Leibnizian formalism we now use collapses this infinite recursion into a single character $\left(\frac{dx}{dt}\right)$ or the integral sign) because one can't actually keep writing operations forever. This makes the character manipulation of calculus even less iconic of the corresponding physical referent.

"So the reference of an operation expressed in calculus is in effect doubly-encrypted. Yes, we've evolved mental capacities well-suited to the manipulation of physical objects, so of course this is difficult. But *math is a form of 'encryption,'* not merely representation, and decryption is an intrinsically difficult process because of the combinatorial challenges it presents. This is why encryption works to make the referential content of communications difficult to recover. My point is that this is *intrinsic to what math is*, irrespective of our evolved capacities. It is difficult for precisely the same reason that deciphering a coded message is difficult.

"What surprises me is that we all know that mathematical equations are encrypted messages, for which you need to know the key if you want to crack the code and know what is represented. Nevertheless, we wonder why higher math is difficult to teach, and often blame the educational system or bad teachers. I think that it is similarly a bit misplaced to blame evolution." (Personal communication with the author, July 11, 2013.)

9   Bilalić et al. 2008.

10 Geary 2011. See also the landmark documentary *A Private Universe*, available at http://www.learner.org/resources/series28.html?pop=yes&pid=9, which led to much research into misconceptions in understanding science.

11 Alan Schoenfeld (1992) notes that in his collection of more than a hundred "videotapes of college and high school students working unfamiliar problems, roughly sixty percent of the solution attempts are of the 'read, make a decision quickly, and pursue that direction come hell or high water' variety." You could characterize this as focused thinking at its worst.

12 Goldacre 2010.

13 Gerardi et al. 2013.

14 Hemispheric differences may sometimes be important, but again, claims in this area should be taken with caution. Norman Cook says it best when he notes:

"Many discussions in the 1970s went well beyond the facts—as hemisphere differ-ences were invoked to explain, in one fell swoop, all of the puzzles of human psychology, including the subconscious mind, creativity, and parapsychological phenomena—but the inevitable backlash was also exaggerated" (Cook 2002, p. 9).

15 Demaree et al. 2005; Gainotti 2012.

16 McGilchrist 2010; Mihov et al. 2010.

17 Nielsen et al. 2013.

18 A differing layout of this problem was provided in de Bono 1970—that was the inspiration for the problem outlined here. De Bono's classic book contains a wealth of such insightful problems and is well worth reading.

19 Immordino-Yang et al. 2012.

20 Although I'm speaking of lobbing between the focused and diffuse modes, there appears to be an analogous lobbing process of information back and forth be-tween the hemispheres. We can get some sense of how information might flow back and forth between the hemispheres in humans by looking at studies of chicks. Learning not to peck a bitter bead involves a complex back and forth processing of the memory traces between the hemispheres over a number of hours (Güntürkün 2003).

   Anke Bouma observes, "An observed pattern of laterality does not mean that the same hemisphere is superior for all of the processing stages required by a particular task. There are indications that the [right hemisphere] may be domi-nant for one stage of processing, while the [left hemisphere] may be dominant for another processing stage. The relative difficulty of a particular processing stage seems to determine which hemisphere is superior for a particular task" (Bouma 1990, p. 86).

21 Just move the coins as shown—do you see how the new triangle will point down?

Chapter 3: Learning Is Creating: Lessons from Thomas Edison's Frying Pan

1 The cerebral distance model developed by Marcel Kinsbourne and Merrill His-cock (1983) hypothesizes that concurrent tasks will interfere more with one another the closer together the two tasks are processed in the brain. Two simul-

taneous tasks using the same hemisphere and particularly the same area of the brain can really mess things up (Bouma 1990, p. 122). Perhaps the diffuse mode may be more capable of handling several tasks at once because of the unfocused nature of diffuse processes.

2 Rocke 2010, p. 316, citing Gruber 1981.

3 Ibid., pp. 3–4.

4 Kaufman et al. 2010, in particular the disinhibition hypothesis on pp. 222–224; Takeuchi et al. 2012.

5 In attempting to track down the provenance of this legend, I corresponded with Leonard DeGraaf, an archivist with the Thomas Edison National Historical Park. He noted, "I have heard the story of Edison and the ball bearings but have never seen any documentation that would confirm it. I'm also not sure about the story's origin. This may be one of those anecdotes that had some basis in reality but became part of the Edison mythology."

6 Dalí 1948, p. 36.

7 Gabora and Ranjan 2013, p. 19.

8 Christopher Lee Niebauer and Garvey 2004. Niebauer refers to the distinction between *object* and *meta-level* thinking. The third, paradoxical error in the sentence, incidentally, is that there is no third error.

9 Kapur and Bielczyc 2012, contains an excellent review on the importance of failure in problem solving.

10 For a nice discussion of the many variations of what Edison actually might have said or written, see http://quoteinvestigator.com/2012/07/31/edison-lot-results/

11 Andrews-Hanna 2012; Raichle and Snyder 2007.

12 Doug Rohrer and Harold Pashler (2010, p. 406) note: ". . . recent analysis of the temporal dynamics of learning show that learning is most durable when study time is distributed over much greater periods of time than is customary in educational settings." How this relates to alternation between focused and resting state networks is an important topic for future research. See Immordino-Yang et al. 2012. In other words, what I've described is a reasonable supposition for what occurs while we learn, but needs to be borne out by further research.

13 Baumeister and Tierney 2011.

14 I want to make it clear that these are only my "best guess" ideas about what might promote diffuse-mode thinking, based on where people seem to get many of their most creative, "aha!" insights.

15 Bilalić et al. 2008.

16 Nakano et al. 2012.

17 Kounios and Beeman 2009, p. 212.

18 Dijksterhuis et al. 2006.

19 Short-term memory is the activated information that is not actively rehearsed. Working memory is the subset of short-term memory information that is the focus of attention and active processing (Baddeley et al. 2009).

20 Cowan 2001.

21 If you're interested in the neural geography underlying all of this, it looks like both long-term memory and working memory use overlapping regions in the frontal and parietal lobes. But the medial temporal lobe is used only for long-term memory—not working memory. See Guida et al. 2012, pp. 225–226, and Dudai 2004.

22 Baddeley et al. 2009, pp. 71–73; Carpenter et al. 2012. *Spaced repetition* is also known as *distributed practice*. Dunlosky et al. 2013, sec. 9, provides an excellent review of distributed practice. Unfortunately, as noted in Rohrer and Pashler 2007, many educators, particularly in mathematics, believe overlearning is a good way to boost long-term retention—hence many similar problems are assigned that ultimately devolve to make-work with little long-term benefit.

23 Xie et al. 2013.

24 Stickgold and Ellenbogen 2008.

25 Ji and Wilson 2006; Oudiette et al. 2011.

26 Ellenbogen et al. 2007. The diffuse mode may also be related to low latent inhibition—that is, being rather absentminded and easily distractable (Carson et al. 2003). There's creative hope for those of us who tend to switch thoughts in the middle of a sentence!

27 Erlacher and Schredl 2010.

28 Wamsley et al. 2010.

Chapter 4: Chunking and Avoiding Illusions of Competence: The Keys to Becoming an "Equation Whisperer"

1 Luria 1968.

2 Beilock 2010, pp. 151–154.

3 Children learn through focused attention, but they also use the diffuse mode, with little executive control, to learn even when they are *not* paying focused attention (Thompson-Schill et al. 2009). In other words, it seems that children don't need to use the focused mode as much as adults do when learning a new language, which may be why it's easier for young children to pick up a new language. But at least some focused learning appears necessary to acquire a new language beyond early childhood.

4 Guida et al. 2012, sec. 8. Recently, Xin Jin, Fatuel Tecuapetla, and Rui Costa revealed how neurons in the basal ganglia play an important role in signaling the concatenation of individual elements into a behavioral sequence—the essence of chunking (Jin et al. 2014). Rui Costa has received a 2 million euro grant to study the mechanism of chunking—his unfolding research will bear watching.

5 Brent and Felder 2012; Sweller et al. 2011, chap. 8.

6 Alessandro Guida and colleagues (2012, p. 235) noted that chunk creation apparently relies initially on working memory, which is in the prefrontal areas, and

results from focused attention, which helps binds chunks. These chunks also begin to reside, with developing expertise, in long-term memory related to the parietal regions. A very different aspect of memory involves neural oscillatory rhythms, which help bind perceptual and contextual information from many areas of the brain (Nyhus and Curran 2010). See Cho et al. 2012 for an imaging study of the development of retrieval fluency in arithmetic problem solving in children.

7 Baddeley et al. 2009, chap. 6; Cree and McRae 2003.

8 Baddeley et al. 2009, pp. 101–104.

9 The "big picture" I'm referring to can be thought of as a cognitive template. See Guida et al. 2012, in particular sec. 3.1. Templates arising from the study of math and science would naturally tend to be more amorphous than those arising from the crisp outlines of chess. Chunks, Guida notes, can be built very quickly, but templates, which involve functional reorganization, take time—at least five weeks or more (Guida et al. 2012). See also the discussion of schemata in Cooper and Sweller 1987; Mastascusa et al. 2011, pp. 23–43. Also useful in understanding these ideas related to developing expertise is the discussion in Bransford et al. 2000, chap. 2. Prior knowledge can be helpful in learning something new and related—but prior knowledge can also act as a hindrance, as it can make it more difficult to make changes in schemata. This is particularly noticeable with students' erroneous embedded beliefs about basic concepts in physics, which are notoriously resistant to change (Hake 1998; Halloun and Hestenes 1985). As Paul Pintrich and colleagues (1993, p. 170) note: "a paradox exists for the learner; on the one hand, current conceptions potentially constitute momentum that resists conceptual change, but they also provide frameworks that the learner can use to interpret and understand new, potentially conflicting information."

10 Geary et al. 2008, pages 4-6 through 4-7; Karpicke 2012; Karpicke et al. 2009; Karpicke and Grimaldi 2012; Kornell et al. 2009; Roediger and Karpicke 2006. For reviews, see McDaniel and Callender 2008; Roediger and Butler 2011.

11 Karpicke et al. 2009, p. 471. See also the Dunning-Kruger effect, where incompetent people mistakenly note their ability higher than they should. Dunning et al. 2003; Kruger and Dunning 1999; Ehrlinger et al. 2008; Bursonet et al. 2006.

12 Baddeley et al. 2009, p. 111.

13 Dunlosky et al. 2013, sec. 4.

14 Longcamp et al. 2008.

15 Dunlosky et al. 2013, sec. 7.

16 See in particular Guida et al. 2012, which notes how experts learn to use long-term memory to expand their working memory. See also Geary et al. 2008, 4-5, which observes, "Working-memory capacity limits mathematical performance, but practice can overcome this limitation by achieving automaticity."

17 The solution to the anagram is "Madame Curie." Attributed to Meyran Kraus, http://www.fun-with-words.com/anag_names.html.

18 Jeffrey Karpicke and colleagues (2009) suggested the relationship between illusions of competence in learning and the difficulty of anagrams when you see the solution as opposed to when you don't see the solution.

19 Henry Roediger and Mary Pyc (2012, p. 243) note: "Professors in schools of education and teachers often worry about creativity in students, a laudable goal. The techniques we advocate show improvements in basic learning and retention of concepts and facts, and some people have criticized this approach as emphasizing "rote learning" or "pure memorization" rather than creative synthesis. Shouldn't education be about fostering a sense of wonder, discovery, and creativity in children? The answer to the question is yes, of course, but we would argue that a strong knowledge base is a prerequisite to being creative in a particular domain. A student is unlikely to make creative discoveries in any subject without a comprehensive set of facts and concepts at his or her command. There is no necessary conflict in learning concepts and facts and in thinking creatively; the two are symbiotic."

20 Geary 2005, chap. 6; Johnson 2010.

21 Johnson 2010, p. 123.

22 Simonton 2004, p. 112.

23 This is my own rephrasing of a common sentiment in science. Santiago Ramón y Cajal cited Duclaux in noting, "Chance smiles not on those who want it, but rather on those who deserve it." Cajal went on to note, "In science as in the lottery, luck favors he who wagers the most—that is, by another analogy, the one who is tilling constantly the ground in his garden" (Ramón y Cajal 1999, pp. 67-68). Louis Pasteur noted, "In the fields of observation chance favors only the prepared mind." Related expressions include the Latin-based proverb "Fortune favors the bold" and the British Special Air Service motto: "Who dares wins."

24 Kounios and Beeman 2009 [1897]; Ramón y Cajal 1999, p. 5.

25 Rocke 2010.

26 Thurston, 1990, p. 846–847.

27 See the foundational work of Karl Anders Ericsson on development of expertise (e.g., Ericsson 2009). For insightful popular approaches related to the development of talent, see Coyle 2009; Greene 2012; Leonard 1991.

28 Karpicke and Blunt 2011a; Karpicke and Blunt 2011b. For further information, see also Guida et al. 2012, p. 239.

29 Of interest is that left hemisphere prefrontal regions appear active during the encoding phase of memorization, while right hemisphere regions are activated during retrieval. This has been reported by many groups using a great variety of imaging techniques (Cook 2002, p. 37). Is it possible that retrieving memorized materials creates the beginnings of diffuse-mode concept mapping-like connections? See also Geary et al. 2008, 4-6 to 4-7.

30 There are, of course, caveats here. For example, what if a student is asked to recall material to determine what belongs on a concept map? There are also undoubtedly disciplinary differences. Some subjects, such as those involving communica-

tion processes in biological cells, inherently lend themselves more readily to "concept map" approaches in understanding key ideas.

31 Brown et al. 1989.

32 Johnson 2010, p. 110.

33 Baddeley et al. 2009, chap. 8.

34 Ken Koedinger, a professor of human-computer interaction and psychology at Carnegie Mellon University, notes, "To maximize retention of material, it's best to start out by exposing the student to the information at short intervals, gradually lengthening the amount of time between encounters. Different types of information—abstract concepts versus concrete facts, for example—require different schedules of exposure" (quoted in Paul 2012).

35 Dunlosky et al. 2013, sec. 10; Roediger and Pyc 2012; Taylor and Rohrer 2010.

36 Rohrer and Pashler 2007.

37 It appears that "mass practice" techniques of presenting the material provide an illusion of competence in teaching. Students appear to learn quickly, but as studies have shown, they forget quickly as well. Roediger and Pyc (2012, p. 244) note: "These outcomes show why teachers and students can be fooled into using strategies that are inefficient in the long run. When we learn we are so focused on how we are learning, we like to adopt strategies that make learning easy and quick. Blocked or massed practice does this. For better retention in the long run, however, we should use spaced and interleaved practice, but while we are learning this procedure seems more arduous. Interleaving makes initial learning more difficult, but is more desirable because long term retention is better."

38 Rohrer et al. 2013.

39 Doug Rohrer and Harold Pashler (2010, p. 406) observe: ". . . the interleaving of different types of practice problems (which is quite rare in math and science texts) markedly improves learning."

40 Personal communication with the author, August 20, 2013. See also Carey 2012.

41 Longcamp et al. 2008.

42 For examples, see http://usefulshortcuts.com/alt-codes.

Chapter 5: Preventing Procrastination: Enlisting Your Habits ("Zombies") as Helpers

1 Emsley 2005, p. 103.

2 Chu and Choi 2005; Graham 2005; Partnoy 2012.

3 Steel (2007, p. 65) notes: "Estimates indicate that 80%–95% of college students engage in procrastination . . . approximately 75% consider themselves procrastinators . . . and almost 50% procrastinate consistently and problematically. The absolute amount of procrastination is considerable, with students reporting that it typically occupies over one third of their daily activities, often enacted through sleeping, playing, or TV watching . . . Furthermore, these percentages appear to be on the rise . . . In addition to being endemic during college, procrastina-

tion is also widespread in the general population, chronically affecting some 15%–20% of adults."

4 Ainslie and Haslam 1992; Steel 2007.

5 Lyons and Beilock 2012.

6 Emmett 2000.

7 See extensive discussion in Duhigg 2012, which in turn cites Weick 1984.

8 Robert Boice (1996, p. 155) noted that procrastination appears to involve a narrowing of the field of consciousness. See also pp. 118–119.

9 Boice 1996, p. 176.

10 Tice and Baumeister 1997.

11 Boice 1996, p. 131.

Chapter 6: Zombies Everywhere: Digging Deeper to Understand the Habit of Procrastination

1 McClain 2011; Wan et al. 2011.

2 Duhigg 2012, p. 274.

3 Steel 2010, p. 190, citing Oaten and Cheng 2006 and Oaten and Cheng 2007.

4 Baumeister and Tierney 2011, pp. 43–51.

5 Steel 2010, citing the original work of Robert Eisenberger, 1992, and others.

6 Ibid., p. 128-130, referring in turn to the work of Gabriele Oettingen.

7 Beilock 2010, pp. 34-35.

8 Ericsson et al. 2007.

9 Boice 1996, pp. 18–22.

10 Paul 2013.

Chapter 7: Chunking versus Choking: How to Increase Your Expertise and Reduce Anxiety

1 One important point is that much of the literature on experts involves individuals who have trained for years to attain their level of expertise. But there are differing levels of experts and expertise. For example, if you know the acronyms FBI and IBM, it's easy to remember the sequence as a chunk of two rather than a disparate grouping of six letters. *But this easy chunking presumes that you are already* an expert, not only with the meaning of FBI and IBM, but with the Roman alphabet itself. Imagine how much more difficult it would be to memorize a Tibetan sequence like this: ཨ་ཧ་ཨ་ཧ། .

When we are learning math and science in the classroom, we are starting with some degree of expertise, and what we are expected to learn through the course of a semester is *nothing* like the vast jump in expertise experienced as a novice becomes a grand master at chess. When you are taking a class in some subject, you're not going to see a dramatic neural difference occurring in one semester, similar to the dramatic difference between a novice and a grandmaster. But there is some indication that neural differences in how you process the material can show up even in a period of a few weeks (Guida et al. 2012). More specifically, Guida and colleagues note that experts preferentially make use of the temporal regions, which are crucial for long-term memory (2012, p. 239). In other words, when we steer students away from building structures in long-term memory, we are making it more difficult for them to acquire expertise. Of course, concentration on memorization alone without creative application is also a problem. Again—any teaching method alone can be misused; variety (not to mention competence) is the spice of life!

2   We've talked about interleaving the study of different techniques while you are studying a topic. But what about interleaving the study of completely different subjects? Unfortunately, there's no research literature available on that as yet (Roediger and Pyc 2012, p. 244), so what I'm suggesting about varying what you are studying is simply common sense and common practice. This will be an interesting area to watch for future research.

3   Kalbfleisch 2004.

4   Guida and colleagues (2012, pp. 236–237) note that chunks in working memory and therefore in long-term memory (LTM) "get larger with practice and expertise . . . the chunks get also richer because more LTM knowledge is associated with each one of them. Moreover, several LTM chunks can become linked to knowledge. And eventually, if an individual becomes an expert, the presence of these links between several chunks can result in the creation of high-level hierarchical chunks. . . . For example, in the game of chess, templates can link to '. . . plans, moves, strategical and tactical concepts, as well as other templates'. . . . We suggest that the functional reorganization of the brain can be detected in expertise acquisition when LTM chunks and knowledge structures exist and are effective in the domain of expertise."

5   Duke et al. 2009.

6   For a good review of the circumstances when deliberate practice is most effective, see Pachman et al. 2013.

7   Roediger and Karpicke 2006, p. 199.

8   Wan et al. 2011. This study sought to define the neural circuits responsible for rapid (within two seconds) intuitive generation of the best next move in spot games of shogi, an extraordinarily complex game of strategy. The part of the brain associated with quick, implicit, unconscious habit (the *precuneus-caudate* circuit) appeared central to the rapid generation of the best next move in professional players. See also McClain 2011.

9   Charness et al. 2005.

10 Karpicke et al. 2009; McDaniel and Callender 2008.

11 Fischer and Bidell 2006, pp. 363–370.

12 Roediger and Karpicke 2006, citing William James's *Principles of Psychology*.

13 Beilock 2010, pp. 54–57.

14 Karpicke and Blunt, 2011b; Mastascusa et al. 2011, chap. 6; Pyc and Rawson 2010; Roediger and Karpicke 2006; Rohrer and Pashler 2010. John Dunlosky and col-

leagues, in their in-depth review of various learning techniques (2013), rate practice testing as having high utility because of its effectiveness, broad applicability, and ease of use. See also Pennebaker et al 2013.

15 Keresztes et al. 2013 provides evidence that testing promotes long-term learning via stabilizing activation patterns in a large network of brain areas.

16 Pashler et al. 2005.

17 Dunlosky et al. 2013, sec. 8; Karpicke and Roediger 2008; Roediger and Karpicke 2006.

Chapter 8: Tools, Tips, and Tricks

1 Allen 2001, pp. 85, 86.

2 Steel 2010, p. 182.

3 Beilock 2010, pp. 162–165; Chiesa and Serretti 2009; Lutz et al. 2008.

4 For those with an interest, please see the resources listed at the Association for Contemplative Mind in Higher Education, http://www.acmhe.org/.

5 Boice 1996, p. 59.

6 Ferriss 2010, p. 485.

7 Ibid., p. 487.

8 Fiore 2007, p. 44.

9 Scullin and McDaniel 2010.

10 Newport 2012; Newport 2006.

11 Fiore 2007, p. 82.

12 Baddeley et al. 2009, pp. 378–379.

Chapter 9: Procrastination Zombie Wrap-Up

1 Johansson 2012, chap. 7.

2 Boice 1996, p. 120; Fiore 2007 chap. 6.

3 Ibid., p. 125.

4 Amabile et al. 2002; Baer and Oldham 2006; Boice 1996, p. 66.

5 Rohrer, et al. (in press).

6 Chi et al. 1981.

7 Noesner 2010.

8 Newport 2012, particularly chap. 1 ("Rule #1").

9 Nakano et al. 2012.

10 Duhigg 2012, p. 137.

11 Newport 2012.

12 See Edelman 2012 for many such ideas.

Chapter 10: Enhancing Your Memory

1 Eleanor Maguire and colleagues (2003) studied individuals renowned for outstanding memory feats in forums such as the World Memory Championships.

"Using neuropsychological measures, as well as structural and functional brain imaging," they found "superior memory was not driven by exceptional intellectual ability or structural brain differences. Rather, [they] found that superior memorizers used a spatial learning strategy, engaging brain regions such as the hippocampus that are critical for memory and for spatial memory in particular."

Tony Buzan has done much to bring the importance of memory techniques to the popular eye. His book *Use Your Perfect Memory* (Buzan, 1991) provides further information about some popular techniques.

2 Eleanor Maguire and colleagues (2003) note that memory techniques are often regarded as being too complicated to use, but some techniques, such as the memory palace, can indeed be very natural and helpful in allowing us to remember information that is important to us.

3 Cai et al. 2013; Foer 2011. Denise Cai and colleagues' work indicates that specialization in one hemisphere (often the left) for language is accompanied by similar specialization in the other hemisphere for visuospatial capabilities. Specialization of a function in one hemisphere, in other words, appears to cause specialization of the other function in the other hemisphere.

4 Ross and Lawrence 1968.

5 Baddeley et al. 2009, pp. 363–365.

6 http://www.ted.com/talks/joshua_foer_feats_of_memory_anyone_can_do.html.

7 http://www.skillstoolbox.com/career-and-education-skills/learning-skills/memory-skills/mnemonics/applications-of-mnemonic-systems/how-to-memorize-formulas/.

8 A sense of the importance of spatial reasoning is provided in Kell et al. 2013.

Chapter 11: More Memory Tips

1 Two sources of information related to metaphor in late-nineteenth-century physics are Cat 2001 and Lützen 2005. For metaphor in chemistry and more broadly throughout science, see Rocke 2010, in particular chap. 11. See also Gentner and Jeziorski 1993. Imagery and visualization are beyond the scope of any single book—see, for example, the *Journal of Mental Imagery*.

2 As leading mathematical modeler Emanuel Derman notes: "Theories describe and deal with the world on its own terms and must stand on their own two feet. Models stand on someone else's feet. They are metaphors that compare the object of their attention to something else that it resembles. Resemblance is always partial, and so models necessarily simplify things and reduce the dimensions of the world. . . . In a nutshell, theories tell you what something is; models tell you merely what something is like" (Derman 2011, p. 6).

3 Solomon 1994.

4 Rocke 2010, p. xvi.

5 Ibid., p. 287, citing *Berichte der Durstigen Chemischen Gesellschaft* (1886), p. 3536. This was a mock issue of the nonexistent "durstigen" (thirsty) Chemical Society.

The parody was sent to the subscribers of the *Berichte der deutschen chemischen Gesellschaft* and is virtually impossible to find today, since it was actually a spurious issue.

6 Rawson and Dunlosky 2011.

7 Dunlosky et al. 2013; Roediger and Pyc 2012. In a review of student flash card use, Kathryn Wissman and colleagues (2012, p. 568) observed: "students understand the benefits of practising to higher criterion levels (amount of practice) but do not typically implement or understand the benefits of practising with longer lags (timing of practice)."

8 Morris et al. 2005.

9 Baddeley et al. 2009, pp. 207–209.

10 In this book, you might think I've discussed all of the components of the SQ3R for study (sometimes SQ4R—for Survey, Question, Read, Recite, Review and wRite). So you might ask why I haven't explored this method further in the text. The SQ3R was developed by psychologist Francis Pleasant Robinson as a general study tool. Central to the study of math and science is problem solving—the SQ3R approach simply doesn't lend itself to this. I'm not the only one to notice. As physics professor Ronald Aaron and his son Robin Aaron note in *Improve Your Physics Grade*, ". . . one Psychology text suggests studying by the SQ3R method. . . . For effective note taking in class it suggests the LISAN approach. . . . Do you believe that such approaches can help you? Do you believe in Santa Claus? The Easter Bunny?" (Aaron and Aaron 1984, p. 2).

11 Curiously, it appears very little work has been done in this area—what little is available seems to simply affirm that writing things out by hand helps us assimilate information better than typing. See Rivard and Straw 2000; Smoker et al. 2009; Velay and Longcamp 2012.

12 Cassilhas et al. 2012; Nagamatsu et al. 2013; van Praag et al. 1999.

13 Guida et al. 2012, p. 230; Leutner et al. 2009.

14 Levin et al. 1992 describes how students who use mnemonics outperform students who apply contextual and free learning styles.

15 Guida et al. 2012 points out that training in memory techniques can speed up the process of acquiring chunks and knowledge structures, thus helping people become experts more rapidly by allowing them to use part of their long-term memory as working memory.

16 Baddeley et al. 2009, pp. 376–377, citing research by Helga and Tony Noice (2007).

Chapter 12: Learning to Appreciate Your Talent

1 Jin et al. 2014.

2 Partnoy 2012, p. 73. Partnoy goes on to note: "Sometimes having an understanding of precisely what we are doing unconsciously can kill the natural spontaneity. If we are too self-conscious, we will impede our instincts when we need them. Yet if we aren't self-conscious at all, we will never improve on our instincts. The chal-

lenge during a period of seconds is to be aware of the factors that go into our decisions . . . but not to be so aware of them that they are stilted and ineffectual" (Partnoy 2012, p. 111).

3 Partnoy 2012, p. 72, citing Klein 1999.

4 Klein 1999, p. 150, citing Klein and Klein 1981. But note the small sample size in Klein and Klein 1981.

5 Mauro Pesenti and colleagues (2001, p. 103) note, "We demonstrated that calculation expertise was not due to increased activity of processes that exist in non-experts; rather, the expert and the non-experts used different brain areas for calculation. We found that the expert could switch between short-term effort-requiring storage strategies and highly efficient episodic memory encoding and retrieval, a process that was sustained by right prefrontal and medial temporal areas."

Already in 1899 brilliant psychologist William James wrote, in his classic *Talks to Teachers on Psychology*: "You now see why 'cramming' must be so poor a mode of study. Cramming seeks to stamp things in by intense application immediately before the ordeal. But a thing thus learned can form but few associations. On the other hand, the same thing recurring on different days, in different contexts, read, recited on, referred to again and again, related to other things and reviewed, gets well wrought into the mental structure. This is the reason why you should enforce on your pupils habits of continuous application" (William 2008, [1899], p. 73).

6 In a classic study, William Chase and Herbert Simon (1973) found that the intuitive generation of next moves by chess experts is based on the superior, quick perception of patterns that has been achieved through practice. Fernand Gobet and colleagues (2001, p. 236) define a chunk as "a collection of elements having strong associations with one another, but weak associations with elements within other chunks."

7 Amidzic et al. 2001; Elo 1978; Simon 1974. A figure of 300,000 chunks was cited by Gobet and Simon 2000.

8 Gobet 2005. Gobet goes on to note that expertise in one domain doesn't transfer to another. That's true—certainly if you learned Spanish, it's not going to help you when you go to order sauerkraut in Germany. But the metaskills are important. If you learn how to learn a language, you can pick up a second language more easily.

That, again, is where developing an expertise in something like chess can be quite valuable—it provides a set of neural structures that are similar to those you need when learning math and science. Even if the neural structures are as simple as *you need to internalize the rules of the game*—that's a valuable insight.

9 Beilock 2010, pp. 77–78; White and Shah 2006.

10 Indeed, there is modest support for this type of finding in the research literature. See Simonton 2009.

11 Carson et al. 2003; Ellenbogen et al. 2007; White and Shah 2011.

12 Merim Bilalić and colleagues (2007) point out that some players with an IQ of
   between 108 and 116 fell into the elite player group by virtue of their extra practice. The elite group had an average IQ of 130. See also Duckworth and Seligman 2005.

   Nobel Prize winner Richard Feynman liked to tout his relatively low IQ score
   of 125 as evidence that you could go pretty far whatever tests might indicate
   about your intelligence. Feynman clearly had natural smarts, but even as a
   youngster he was practicing obsessively in developing his mathematical and
   physical knowledge and intuition (Gleick 1992).
13 Klingberg 2008.
14 Silverman 2012.
15 Felder 1988. See also Justin Kruger and David Dunning (1999), who note "the
   miscalibration of the incompetent stems from an error about the self, where
   the miscalibration of the highly competent stems from an error about others."

Chapter 13: Sculpting Your Brain

1 DeFelipe 2002.
2 Ramón y Cajal 1937, 309.
3 Ramón y Cajal 1999 [1897], pp. xv–xvi; Ramón y Cajal 1937, p. 278.
4 Ramón y Cajal 1937, 154.
5 Fields 2008; Giedd 2004; Spear 2013.
6 Ramón y Cajal 1999 [1897].
7 Bengtsson et al. 2005; Spear 2013.
8 Cajal could clearly plan well—witness his construction of the cannon. But he
   couldn't seem to make the connection with the bigger picture consequences of
   his actions. Taken up with the exciting task of blowing up a neighbor's gate, for
   example, he couldn't make the obvious prediction that he would be in deep
   trouble as a consequence. See Shannon et al. 2011, with their intriguing finding
   that functional connectivity in troubled teens connects the dorsolateral premotor cortex to the default-mode network ("a constellation of brain areas associated with spontaneous, unconstrained, self-referential cognition" p. 11241). As
   troubled teens mature and their behavior improves, the dorsolateral premotor
   cortex instead appears to begin connecting with the attention and control networks.
9 Bengtsson et al. 2005; Spear 2013; Thomas and Baker 2013. As Cibu Thomas and
   colleagues note (p. 226), "the evidence from animal studies suggests that the
   large-scale organization of axons and dendrites is very stable and experience-
   dependent structural plasticity in the adult brain occurs locally and is transient."
   In other words, we can make modest changes in our brain, but we can't indulge in
   wholesale rewiring. This is all commonsense stuff. For a terrific popular book
   on brain plasticity, see Doidge 2007. The best technical approach to this topic is
   Shaw and McEachern 2001. It is fitting that Cajal's own work is now gaining recognition as foundational in our understanding of brain plasticity (DeFelipe 2006).

10 Ramón y Cajal 1937, p. 58.
11 Ibid., pp. 58, 131. The ability to grasp the key ideas—the gist of the problems—appears to be more important than verbatim ability to memorize. Verbatim as opposed to "gist" memories seem to be encoded differently. See Geary et al. 2008, 4–9.
12 DeFelipe 2002.
13 Ramón y Cajal 1937, p. 59.
14 Root-Bernstein and Root-Bernstein 1999, pp. 88–89.
15 Bransford et al. 2000, chap. 3; Mastascusa et al. 2011, chaps. 9–10.
16 Fauconnier and Turner 2002.
17 Mastascusa et al. 2011, p. 165.
18 Gentner and Jeziorski 1993.

Chapter 14: Developing the Mind's Eye through Equation Poems

1 Plath 1971, p. 34.
2 Feynman 2001, p. 54.
3 Feynman 1965, 2010.
4 This section is based on the wonderful paper by Prentis (1996).
5 Excerpts from the song "Mandelbrot Set," © Jonathan Coulton, by kind permission of Jonathan Coulton. Lyrics excerpted from song fully given at http://www.jonathancoulton.com/wiki/Mandelbrot_Set/Lyrics.
6 Prentis 1996.
7 Cannon 1949, p. xiii; Ramón y Cajal 1937, p. 363. In a related vein, see Javier DeFelipe's extraordinary *Butterflies of the Soul*, which contains some of the beautiful illustrations produced in the early days of research in neuroscience (DeFelipe 2010).
8 Mastascusa et al. 2011, p. 165.
9 Keller 1984, p. 117.
10 See discussions of elaborative interrogation and self-explanation in Dunlosky et al. 2013.
11 http://www.youtube.com/watch?v=FrNqSLPaZLc.
12 http://www.reddit.com/r/explainlikeimfive.
13 See also endnote 8 from chapter 12.
14 Mastascusa et al. 2011, chaps. 9–10.
15 Foerde et al. 2006; Paul 2013.

Chapter 15: Renaissance Learning

1 Colvin 2008; Coyle 2009; Gladwell 2008.
2 Deslauriers et al. 2011; Felder et al. 1998; Hake 1998; Mitra et al. 2005; President's Council of Advisors on Science and Technology, 2012.
3 Ramón y Cajal 1999 [1897].
4 Kamkwamba and Mealer 2009.

5   Pert 1997, p. 33.

6   McCord 1978. See Armstrong 2012 for an extensive discussion of this and related studies. Manu Kapur and Katerine Bielaczyc (2012) indicate that less heavy-handed guidance by instructors may result in counterintuitive improvement in student performance.

7   Oakley et al. 2003.

8   See Armstrong 2012 and references therein.

9   Oakley 2013.

Chapter 16: Avoiding Overconfidence: The Power of Teamwork

1   Schutz 2005. "Fred" is a hypothetical amalgam of typical traits of "broad-perspective perceptual disorder of the right hemisphere."

2   McGilchrist 2010 provides a comprehensive description supporting the differences in hemispheric function, while Efron 1990, although dated, provides an excellent cautionary note about problems in hemispheric research. See also Nielsen et al. 2013; Jeff Anderson, M.D., Ph.D., who was involved in the study, notes, "It's absolutely true that some brain functions occur in one or the other side of the brain. Language tends to be on the left, attention more on the right. But people don't tend to have a stronger left- or right-sided brain network. It seems to be determined more connection by connection" (University of Utah Health Care Office of Public Affairs 2013).

3   McGilchrist 2010, pp. 192–194, 203.

4   Houdé and Tzourio-Mazoyer 2003. Houdé 2002, p. 341 notes, "our neuroimaging results demonstrate the direct involvement, in neurologically intact subjects, of a right ventromedial prefrontal area in the making of logical consciousness, that is, in what puts the mind on 'the logical track,' where it can implement the instruments of deduction. . . . Hence, the right ventromedial prefrontal cortex may be the emotional component of the brain's error correction device. More exactly, this area may correspond to the self-feeling device that detects the conditions under which logical reasoning errors are likely to occur."

5   See Stephen Christman and colleagues 2008, p. 403, who note that "the left hemisphere maintains our current beliefs while the right hemisphere evaluates and updates those beliefs when appropriate. Belief evaluation is thus dependent on interhemispheric interaction."

6   Ramachandran 1999, p. 136.

7   Gazzaniga 2000; Gazzaniga et al. 1996.

8   Feynman 1985, p. 341. Originally given in his 1974 Caltech commencement address.

9   Feynman 1985, pp. 132–133.

10   As Alan Baddeley and colleagues (2009, pp. 148–149) note: "We are not lacking in ways of defending ourselves against challenges to our self-esteem. We readily accept praise but tend to be skeptical of criticism, often attributing criticism to prejudice on the part of the critic. We are inclined to take credit for success when

it occurs but deny responsibility for failure. If this stratagem fails, we are rather good at selectively forgetting failure and remembering success and praise." (References omitted.)

11 Granovetter 1983; Granovetter 1973.
12 Ellis et al. 2003.
13 Beilock 2010, p. 34.
14 Arum and Roksa 2010, p. 120.

Chapter 17: Test Taking

1 Visit Dr. Felder's website at http://www4.ncsu.edu/unity/lockers/users/f/felder/public/ for an enormous assortment of useful information on learning in the STEM disciplines.
2 Felder 1999. Used by permission of Dr. Richard Felder and *Chemical Engineering Education.*
3 For food for thought along these lines, see McClain 2011 and the work of the researchers McClain cites.
4 Beilock 2010, pp. 140–141.
5 Mrazek et al. 2013.
6 Beilock (2010, p. 60) notes that "athletes under pressure sometimes try to control their performance in a way that disrupts it. This control, which is often referred to as 'paralysis by analysis,' stems from an overactive prefrontal cortex."
7 Beilock 2010; http://www.sianbeilock.com/.

references

Aaron, R, and RH Aaron. *Improve Your Physics Grade*. New York: Wiley, 1984.

Ainslie, G, and N Haslam. "Self-control." In *Choice over Time*, edited by G Loewenstein and J Elster, 177–212. New York: Russell Sage Foundation, 1992.

Allen, D. *Getting Things Done*. New York: Penguin, 2001.

Amabile, TM, et al. "Creativity under the gun." *Harvard Business Review* 80, 8 (2002): 52.

Amidzic, O, et al. "Pattern of focal γ-bursts in chess players." *Nature* 412 (2001): 603–604.

Andrews-Hanna, JR. "The brain's default network and its adaptive role in internal mentation." *Neuroscientist* 18, 3 (2012): 251–270.

Armstrong, JS. "Natural learning in higher education." In *Encyclopedia of the Sciences of Learning*, 2426–2433. New York: Springer, 2012.

Arum, R, and J Roksa. *Academically Adrift*. Chicago: University of Chicago Press, 2010.

Baddeley, A, et al. *Memory*. New York: Psychology Press, 2009.

Baer, M, and GR Oldham. "The curvilinear relation between experienced creative time pressure and creativity: Moderating effects of openness to experience and support for creativity." *Journal of Applied Psychology* 91, 4 (2006): 963–970.

Baumeister, RF, and J Tierney. *Willpower*. New York: Penguin, 2011.

Beilock, S. *Choke:* New York: Free Press, 2010.

Bengtsson, SL, et al. "Extensive piano practicing has regionally specific effects on white matter development." *Nature Neuroscience* 8, 9 (2005): 1148–1150.

Bilalić, M, et al. "Does chess need intelligence?—A study with young chess players." *Intelligence* 35, 5 (2007): 457–470.

———. "Why good thoughts block better ones: The mechanism of the pernicious Einstellung (set) effect." *Cognition* 108, 3 (2008): 652–661.

Boice, R. *Procrastination and Blocking*. Westport, CT: Praeger, 1996.

Bouma, A. *Lateral Asymmetries and Hemispheric Specialization.* Rockland, MA: Swets & Zeitlinger, 1990.

Bransford, JD, et al. *How People Learn.* Washington, DC: National Academies Press, 2000.

Brent, R, and RM Felder. "Learning by solving solved problems." *Chemical Engineering Education* 46, 1 (2012): 29–30.

Brown, JS, et al. "Situated cognition and the culture of learning." *Educational Researcher* 18, 1 (1989): 32–42.

Burson K, et al. "Skilled or unskilled, but still unaware of it: how perceptions of difficulty drive miscalibration in relative comparisons." *Journal of Personality and Social Psychology* 90, 1 (2006): 60–77.

Buzan, T. *Use Your Perfect Memory.* New York: Penguin, 1991.

Cai, Q, et al. "Complementary hemispheric specialization for language production and visuospatial attention." *PNAS* 110, 4 (2013): E322–E330.

Cannon, DF. *Explorer of the Human Brain.* New York: Schuman, 1949.

Carey, B. "Cognitive science meets pre-algebra." *New York Times*, September 2, 2012; http://www.nytimes.com/2013/09/03/science/cognitive-science-meets-pre-algebra.html?ref=science.

Carpenter, SK, et al. "Using spacing to enhance diverse forms of learning: Review of recent research and implications for instruction." *Educational Psychology Review* 24, 3 (2012): 369–378.

Carson, SH, et al. "Decreased latent inhibition is associated with increased creative achievement in high-functioning individuals." *Journal of Personality and Social Psychology* 85, 3 (2003): 499–506.

Cassilhas, RC, et al. "Spatial memory is improved by aerobic and resistance exercise through divergent molecular mechanisms." *Neuroscience* 202 (2012): 309–17.

Cat, J. "On understanding: Maxwell on the methods of illustration and scientific metaphor." *Studies in History and Philosophy of Science Part B* 32, 3 (2001): 395–441.

Charness, N, et al. "The role of deliberate practice in chess expertise." *Applied Cognitive Psychology* 19, 2 (2005): 151–165.

Chase, WG, and HA Simon. "Perception in chess." *Cognitive Psychology* 4, 1 (1973): 55–81.

Chi, MTH, et al. "Categorization and representation of physics problems by experts and novices." *Cognitive Science* 5, 2 (1981): 121–152.

Chiesa, A, and A Serretti. "Mindfulness-based stress reduction for stress management in healthy people: A review and meta-analysis." *Journal of Alternative Complementary Medicine* 15, 5 (2009): 593–600.

Cho, S, et al. "Hippocampal-prefrontal engagement and dynamic causal interactions in the maturation of children's fact retrieval." *Journal of Cognitive Neuroscience* 24, 9 (2012): 1849–1866.

Christman, SD, et al. "Mixed-handed persons are more easily persuaded and are more gullible: Interhemispheric interaction and belief updating." *Laterality* 13, 5 (2008): 403–426.

Chu, A, and JN Choi. "Rethinking procrastination: Positive effects of 'active' procrastination behavior on attitudes and performance." *Journal of Social Psychology* 145, 3 (2005): 245–264.

Colvin, G. *Talent Is Overrated.* New York: Portfolio, 2008.

Cook, ND. *Tone of Voice and Mind.* Philadelphia: Benjamins, 2002.

———. "Toward a central dogma for psychology." *New Ideas in Psychology* 7, 1 (1989): 1–18.

Cooper, G, and J Sweller. "Effects of schema acquisition and rule automation on mathematical problem-solving transfer." *Journal of Educational Psychology* 79, 4 (1987): 347.

Cowan, N. "The magical number 4 in short-term memory: A reconsideration of mental storage capacity." *Behavioral and Brain Sciences* 24, 1 (2001): 87–114.

Coyle, D. *The Talent Code.* New York: Bantam, 2009.

Cree, GS, and K McRae. "Analyzing the factors underlying the structure and computation of the meaning of chipmunk, cherry, chisel, cheese, and cello (and many other such concrete nouns)." *Journal of Experimental Psychology: General* 132, 2 (2003): 163–200.

Dalí, S. *Fifty Secrets of Magic Craftsmanship.* New York: Dover, 1948 (reprint 1992).

de Bono, E. *Lateral Thinking.* New York: Harper Perennial, 1970.

DeFelipe, J. "Brain plasticity and mental processes: Cajal again." *Nature Reviews Neuroscience* 7, 10 (2006): 811–817.

———. *Cajal's Butterflies of the Soul: Science and Art.* New York: Oxford University Press, 2010.

———. "Sesquicentenary of the birthday of Santiago Ramón y Cajal, the father of modern neuroscience." *Trends in Neurosciences* 25, 9 (2002): 481–484.

Demaree, H, et al. "Brain lateralization of emotional processing: Historical roots and a future incorporating 'dominance.'" *Behavioral and Cognitive Neuroscience Reviews* 4, 1 (2005): 3–20.

Derman, E. *Models. Behaving. Badly.* New York: Free Press, 2011.

Deslauriers, L, et al. "Improved learning in a large-enrollment physics class." *Science* 332, 6031 (2011): 862–864.

Dijksterhuis, A, et al. "On making the right choice: The deliberation-without-attention effect." *Science* 311, 5763 (2006): 1005–1007.

Doidge, N. *The Brain That Changes Itself.* New York: Penguin, 2007.

Drew, C. "Why science majors change their minds (it's just so darn hard)." *New York Times,* November 4, 2011.

Duckworth, AL, and ME Seligman. "Self-discipline outdoes IQ in predicting academic performance of adolescents." *Psychological Science* 16, 12 (2005): 939–944.

Dudai, Y. "The neurobiology of consolidations, or, how stable is the engram?" *Annual Review of Psychology* 55 (2004): 51–86.

Duhigg, C. *The Power of Habit*. New York: Random House, 2012.

Duke, RA, et al. "It's not how much; it's how: Characteristics of practice behavior and retention of performance skills." *Journal of Research in Music Education* 56, 4 (2009): 310–321.

Dunlosky, J, et al. "Improving students' learning with effective learning techniques: Promising directions from cognitive and educational psychology." *Psychological Science in the Public Interest* 14, 1 (2013): 4–58.

Dunning, D, et al. "Why people fail to recognize their own incompetence." *Current Directions in Psychological Science* 12, 3 (2003): 83–87.

Dweck, C. *Mindset*. New York: Random House, 2006.

Edelman, S. *Change Your Thinking with CBT*. New York: Ebury, 2012.

Efron, R. *The Decline and Fall of Hemispheric Specialization*. Hillsdale, NJ: Erlbaum, 1990.

Ehrlinger, J, et al. "Why the unskilled are unaware: Further explorations of (absent) self-insight among the incompetent." *Organizational Behavior and Human Decision Processes* 105, 1 (2008): 98–121.

Eisenberger, R. "Learned industriousness." *Psychological Review* 99, 2 (1992): 248.

Ellenbogen, JM, et al. "Human relational memory requires time and sleep." *PNAS* 104, 18 (2007): 7723–7728.

Ellis, AP, et al. "Team learning: Collectively connecting the dots." *Journal of Applied Psychology* 88, 5 (2003): 821.

Elo, AE. *The Rating of Chessplayers, Past and Present*. London: Batsford, 1978.

Emmett, R. *The Procrastinator's Handbook*. New York: Walker, 2000.

Emsley, J. *The Elements of Murder*. New York: Oxford University Press, 2005.

Ericsson, KA. *Development of Professional Expertise*. New York: Cambridge University Press, 2009.

Ericsson, KA, et al. "The making of an expert." *Harvard Business Review* 85, 7/8 (2007): 114.

Erlacher, D, and M Schredl. "Practicing a motor task in a lucid dream enhances subsequent performance: A pilot study." *The Sport Psychologist* 24, 2 (2010): 157–167.

Fauconnier, G, and M Turner. *The Way We Think*. New York: Basic Books, 2002.

Felder, RM. "Memo to students who have been disappointed with their test grades." *Chemical Engineering Education* 33, 2 (1999): 136–137.

——— "Impostors everywhere." *Chemical Engineering Education* 22, 4 (1988): 168–169.

Felder, RM, et al. "A longitudinal study of engineering student performance and retention. V. Comparisons with traditionally-taught students." *Journal of Engineering Education* 87, 4 (1998): 469–480.

Ferriss, T. *The 4-Hour Body*. New York: Crown, 2010.

Feynman, R. *The Feynman Lectures on Physics Vol. 2*. New York: Addison Wesley, 1965.

———. *"Surely You're Joking, Mr. Feynman."* New York: Norton, 1985.

———. *What Do You Care What Other People Think?* New York: Norton, 2001.

Fields, RD. "White matter in learning, cognition and psychiatric disorders." *Trends in Neurosciences* 31, 7 (2008): 361–370.

Fiore, NA. *The Now Habit.* New York: Penguin, 2007.

Fischer, KW, and TR Bidell. "Dynamic development of action, thought, and emotion." In *Theoretical Models of Human Development: Handbook of Child Psychology,* edited by W Damon and RM Lerner. New York: Wiley, 2006: 313–399.

Foer, J. *Moonwalking with Einstein.* New York: Penguin, 2011.

Foerde, K, et al. "Modulation of competing memory systems by distraction." *Proceedings of the National Academy of the Sciences* 103, 31 (2006): 11778–11783.

Gabora, L, and A Ranjan. "How insight emerges in a distributed, content-addressable memory." In *Neuroscience of Creativity,* edited by O Vartanian et al. Cambridge, MA: MIT Press, 2013: 19–43.

Gainotti, G. "Unconscious processing of emotions and the right hemisphere." *Neuropsychologia* 50, 2 (2012): 205–218.

Gazzaniga, MS. "Cerebral specialization and interhemispheric communication: Does the corpus callosum enable the human condition?" *Brain* 123, 7 (2000): 1293–1326.

Gazzaniga, MS, et al. "Collaboration between the hemispheres of a callosotomy patient: Emerging right hemisphere speech and the left hemisphere interpreter." *Brain* 119, 4 (1996): 1255–1262.

Geary, DC. *The Origin of Mind.* Washington, DC: American Psychological Association, 2005.

———. "Primal brain in the modern classroom." *Scientific American Mind* 22, 4 (2011): 44–49.

Geary, DC, et al. "Task Group Reports of the National Mathematics Advisory Panel; Chapter 4: Report of the Task Group on Learning Processes." 2008. http://www2.ed.gov/about/bdscomm/list/mathpanel/report/learning-processes.pdf.

Gentner, D, and M Jeziorski. "The shift from metaphor to analogy in western science." In *Metaphor and Thought,* edited by A Ortony. 447–480, Cambridge, UK: Cambridge University Press, 1993.

Gerardi, K, et al. "Numerical ability predicts mortgage default." *Proceedings of the National Academy of Sciences* 110, 28 (2013): 11267–11271.

Giedd, JN. "Structural magnetic resonance imaging of the adolescent brain." *Annals of the New York Academy of Sciences* 1021, 1 (2004): 77–85.

Gladwell, M. *Outliers.* New York: Hachette, 2008.

Gleick, J. *Genius.* New York: Pantheon Books, 1992.

Gobet, F. "Chunking models of expertise: Implications for education." *Applied Cognitive Psychology* 19, 2 (2005): 183–204.

Gobet, F, et al. "Chunking mechanisms in human learning." *Trends in Cognitive Sciences* 5, 6 (2001): 236–243.

Gobet, F, and HA Simon. "Five seconds or sixty? Presentation time in expert memory." *Cognitive Science* 24, 4 (2000): 651–682.

Goldacre, B. *Bad Science*. London: Faber & Faber, 2010.

Graham, P. "Good and bad procrastination." 2005. http://paulgraham.com/procrastination.html.

Granovetter, M. "The strength of weak ties: A network theory revisited." *Sociological Theory* 1, 1 (1983): 201–233.

Granovetter, MS. "The strength of weak ties." *American Journal of Sociology* (1973): 1360–1380.

Greene, R. *Mastery*. New York: Viking, 2012.

Gruber, HE. "On the relation between aha experiences and the construction of ideas." *History of Science Cambridge* 19, 1 (1981): 41–59.

Guida, A, et al. "How chunks, long-term working memory and templates offer a cognitive explanation for neuroimaging data on expertise acquisition: A two-stage framework." *Brain and Cognition* 79, 3 (2012): 221–244.

Güntürkün, O. "Hemispheric asymmetry in the visual system of birds." In *The Asymmetrical Brain*, edited by K Hugdahl and RJ Davidson, 3–36. Cambridge, MA: MIT Press, 2003.

Hake, RR. "Interactive-engagement versus traditional methods: A six-thousand-student survey of mechanics test data for introductory physics courses." *American Journal of Physics* 66 (1998): 64–74.

Halloun, IA, and D Hestenes. "The initial knowledge state of college physics students." *American Journal of Physics* 53, 11 (1985): 1043–1055.

Houdé, O. "Consciousness and unconsciousness of logical reasoning errors in the human brain." *Behavioral and Brain Sciences* 25, 3 (2002): 341–341.

Houdé, O, and N Tzourio-Mazoyer. "Neural foundations of logical and mathematical cognition." *Nature Reviews Neuroscience* 4, 6 (2003): 507–513.

Immordino-Yang, MH, et al. "Rest is not idleness: Implications of the brain's default mode for human development and education." *Perspectives on Psychological Science* 7, 4 (2012): 352–364.

James, W. *Principles of Psychology*. New York: Holt, 1890.

———. *Talks to Teachers on Psychology: And to Students on Some of Life's Ideals*. Rockville, MD: ARC Manor, 2008 [1899].

Ji, D, and MA Wilson. "Coordinated memory replay in the visual cortex and hippocampus during sleep." *Nature Neuroscience* 10, 1 (2006): 100–107.

Jin, X. "Basal ganglia subcircuits distinctively encode the parsing and concatenation of action sequences." *Nature Neuroscience* 17 (2014): 423–430.

Johansson, F. *The Click Moment*. New York: Penguin, 2012.

Johnson, S. *Where Good Ideas Come From*. New York: Riverhead, 2010.

Kalbfleisch, ML. "Functional neural anatomy of talent." *The Anatomical Record Part B: The New Anatomist* 277, 1 (2004): 21–36.

Kamkwamba, W, and B Mealer. *The Boy Who Harnessed the Wind*. New York: Morrow, 2009.

Kapur, M, and K Bielczyc. "Designing for productive failure." *Journal of the Learning Sciences* 21, 1 (2012): 45–83.

Karpicke, JD. "Retrieval-based learning: Active retrieval promotes meaningful learning." *Current Directions in Psychological Science* 21, 3 (2012): 157–163.

Karpicke, JD, and JR Blunt. "Response to comment on 'Retrieval practice produces more learning than elaborative studying with concept mapping.'" *Science* 334, 6055 (2011a): 453–453.

———. "Retrieval practice produces more learning than elaborative studying with concept mapping." *Science* 331, 6018 (2011b): 772–775.

Karpicke, JD, et al. "Metacognitive strategies in student learning: Do students practice retrieval when they study on their own?" *Memory* 17, 4 (2009): 471–479.

Karpicke, JD, and PJ Grimaldi. "Retrieval-based learning: A perspective for enhancing meaningful learning." *Educational Psychology Review* 24, 3 (2012): 401–418.

Karpicke, JD, and HL Roediger. "The critical importance of retrieval for learning." *Science* 319, 5865 (2008): 966–968.

Kaufman, AB, et al. "The neurobiological foundation of creative cognition." *Cambridge Handbook of Creativity* (2010): 216–232.

Kell, HJ, et al. "Creativity and technical innovation: Spatial ability's unique role." *Psychological Science* 24, 9 (2013): 1831–1836.

Keller, EF. *A Feeling for the Organism, 10th Aniversary Edition: The Life and Work of Barbara McClintock*. New York: Times Books, 1984.

Keresztes, A, et al. "Testing promotes long-term learning via stabilizing activation patterns in a large network of brain areas." *Cerebral Cortex* (advance access, published June 24, 2013).

Kinsbourne, M, and M Hiscock. "Asymmetries of dual-task performance." In *Cerebral Hemisphere Asymmetry*, edited by JB Hellige, 255–334. New York: Praeger, 1983.

Klein, G. *Sources of Power*. Cambridge, MA: MIT Press, 1999.

Klein, H, and G Klein. "Perceptual/cognitive analysis of proficient cardio-pulmonary resuscitation (CPR) performance." Midwestern Psychological Association Conference, Detroit, MI, 1981.

Klingberg, T. *The Overflowing Brain*. New York: Oxford University Press, 2008.

Kornell, N, et al. "Unsuccessful retrieval attempts enhance subsequent learning." *Journal of Experimental Psychology: Learning, Memory, and Cognition* 35, 4 (2009): 989.

Kounios, J, and M Beeman. "The Aha! moment: The cognitive neuroscience of insight." *Current Directions in Psychological Science* 18, 4 (2009): 210–216.

Kruger, J, and D Dunning. "Unskilled and unaware of it: How difficulties in one's own incompetence lead to inflated self-assessments." *Journal of Personality and Social Psychology* 77, 6 (1999): 1121–1134.

Leonard, G. *Mastery*. New York: Plume, 1991.

Leutner, D, et al. "Cognitive load and science text comprehension: Effects of drawing and mentally imaging text content." *Computers in Human Behavior* 25 (2009): 284–289.

Levin, JR, et al. "Mnemonic vocabulary instruction: Additional effectiveness evidence." *Contemporary Educational Psychology* 17, 2 (1992): 156–174.

Longcamp, M, et al. "Learning through hand- or typewriting influences visual recognition of new graphic shapes: Behavioral and functional imaging evidence." *Journal of Cognitive Neuroscience* 20, 5 (2008): 802–815.

Luria, AR. *The Mind of a Mnemonist*. Translated by L Solotaroff. New York: Basic Books, 1968.

Lutz, A, et al. "Attention regulation and monitoring in meditation." *Trends in Cognitive Sciences* 12, 4 (2008): 163.

Lützen, J. *Mechanistic Images in Geometric Form*. New York: Oxford University Press, 2005.

Lyons, IM, and SL Beilock. "When math hurts: Math anxiety predicts pain network activation in anticipation of doing math." *PLOS ONE* 7, 10 (2012): e48076.

Maguire, EA, et al. "Routes to remembering: The brains behind superior memory." *Nature Neuroscience* 6, 1 (2003): 90–95.

Mangan, BB. "Taking phenomenology seriously: The 'fringe' and its implications for cognitive research." *Consciousness and Cognition* 2, 2 (1993): 89–108.

Mastascusa, EJ, et al. *Effective Instruction for STEM Disciplines*. San Francisco: Jossey-Bass, 2011.

McClain, DL. "Harnessing the brain's right hemisphere to capture many kings." *New York Times*, January 24 (2011). http://www.nytimes.com/2011/01/25/science/25chess.html?_r=0.

McCord, J. "A thirty-year follow-up of treatment effects." *American Psychologist* 33, 3 (1978): 284.

McDaniel, MA, and AA Callender. "Cognition, memory, and education." In *Cognitive Psychology of Memory, Vol. 2 of Learning and Memory*, edited by HL Roediger, 819–843. Oxford, UK: Elsevier, 2008.

McGilchrist, I. *The Master and His Emissary*. New Haven, CT: Yale University Press, 2010.

Mihov, KM, et al. "Hemispheric specialization and creative thinking: A meta-analytic review of lateralization of creativity." *Brain and Cognition* 72, 3 (2010): 442–448.

Mitra, S, et al. "Acquisition of computing literacy on shared public computers: Children and the 'hole in the wall.'" *Australasian Journal of Educational Technology* 21, 3 (2005): 407.

Morris, PE, et al. "Strategies for learning proper names: Expanding retrieval practice, meaning and imagery." *Applied Cognitive Psychology* 19, 6 (2005): 779–798.

Moussa, MN, et al. "Consistency of network modules in resting-state fMRI connectome data." *PLOS ONE* 7, 8 (2012): e49428.

Mrazek, M, et al. "Mindfulness training improves working memory capacity and GRE performance while reducing mind wandering." *Psychological Science* 24, 5 (2013): 776–781.

Nagamatsu, LS, et al. "Physical activity improves verbal and spatial memory in adults with probable mild cognitive impairment: A 6-month randomized controlled trial." *Journal of Aging Research* (2013): 861893.

Nakano, T, et al. "Blink-related momentary activation of the default mode network while viewing videos." *Proceedings of the National Academy of Sciences* 110, 2 (2012): 702–706.

National Survey of Student Engagement. *Promoting Student Learning and Institutional Improvement: Lessons from NSSE at 13.* Bloomington: Indiana University Center for Postsecondary Research, 2012.

Newport, C. *How to Become a Straight-A Student.* New York: Random House, 2006.

———. *So Good They Can't Ignore You.* New York: Business Plus, 2012.

Niebauer, CL, and K Garvey. "Gödel, Escher, and degree of handedness: Differences in interhemispheric interaction predict differences in understanding self-reference." *Laterality: Asymmetries of Body, Brain and Cognition* 9, 1 (2004): 19–34.

Nielsen, JA, et al. "An evaluation of the left-brain vs. right-brain hypothesis with resting state functional connectivity magnetic resonance imaging." *PLOS ONE* 8, 8 (2013).

Noesner, G. *Stalling for Time.* New York: Random House, 2010.

Noice, H, and T Noice. "What studies of actors and acting can tell us about memory and cognitive functioning." *Current Directions in Psychological Science* 15, 1 (2006): 14–18.

Nyhus, E, and T Curran. "Functional role of gamma and theta oscillations in episodic memory." *Neuroscience and Biobehavioral Reviews* 34, 7 (2010): 1023–1035.

Oakley, BA. "Concepts and implications of altruism bias and pathological altruism." *Proceedings of the National Academy of Sciences* 110, Supplement 2 (2013): 10408–10415.

Oakley, B, et al. "Turning student groups into effective teams." *Journal of Student Centered Learning* 2, 1 (2003): 9–34.

Oaten, M, and K Cheng. "Improved self-control: The benefits of a regular program of academic study." *Basic and Applied Social Psychology* 28, 1 (2006): 1–16.

Oaten, M, and K Cheng. "Improvements in self-control from financial monitoring." *Journal of Economic Psychology* 28, 4 (2007): 487–501.

Oettingen, G, et al. "Turning fantasies about positive and negative futures into self-improvement goals." *Motivation and Emotion* 29, 4 (2005): 236–266.

Oettingen, G, and J Thorpe. "Fantasy realization and the bridging of time." In *Judgments over Time: The Interplay of Thoughts, Feelings, and Behaviors,* edited by Sanna, LA and EC Chang, 120–142. New York: Oxford University Press, 2006.

Oudiette, D, et al. "Evidence for the re-enactment of a recently learned behavior during sleepwalking." *PLOS ONE* 6, 3 (2011): e18056.

Pachman, M, et al. "Levels of knowledge and deliberate practice." *Journal of Experimental Psychology* 19, 2 (2013): 108–119.

Partnoy, F. *Wait.* New York: Public Affairs, 2012.

Pashler, H, et al. "When does feedback facilitate learning of words?" *Journal of Experimental Psychology: Learning, Memory, and Cognition* 31, 1 (2005): 3–8.

Paul, AM. "The machines are taking over." *New York Times,* September 14 (2012). http://www.nytimes.com/2012/09/16/magazine/how-computerized-tutors-are-learning-to-teach-humans.html?pagewanted=all.

———. "You'll never learn! Students can't resist multitasking, and it's impairing their memory." *Slate*, May 3 (2013). http://www.slate.com/articles/health_and _science/science/2013/05/multitasking_while_studying_divided_attention _and_technological_gadgets.3.html.

Pennebaker, JW, et al. "Daily online testing in large classes: Boosting college performance while reducing achievement gaps." *PLOS ONE* 8, 11 (2013): e79774.

Pert, CB. *Molecules of Emotion*. New York: Scribner, 1997.

Pesenti, M, et al. "Mental calculation in a prodigy is sustained by right prefrontal and medial temporal areas." *Nature Neuroscience* 4, 1 (2001): 103–108.

Pintrich, PR, et al. "Beyond cold conceptual change: The role of motivational beliefs and classroom contextual factors in the process of conceptual change." *Review of Educational Research* 63, 2 (1993): 167–199.

Plath, S. *The Bell Jar*. New York: Harper Perennial, 1971.

Prentis, JJ. "Equation poems." *American Journal of Physics* 64, 5 (1996): 532–538.

President's Council of Advisors on Science and Technology. *Engage to Excel: Producing One Million Additional College Graduates with Degrees in Science, Technology, Engineering, and Mathematics*. 2012. http://www.whitehouse.gov/sites/default/files/ microsites/ostp/pcast-engage-to-excel-final_feb.pdf

Pyc, MA, and KA Rawson. "Why testing improves memory: Mediator effectiveness hypothesis." *Science* 330, 6002 (2010): 335–335.

Raichle, ME, and AZ Snyder. "A default mode of brain function: A brief history of an evolving idea." *NeuroImage* 37, 4 (2007): 1083–1090.

Ramachandran, VS. *Phantoms in the Brain*. New York: Harper Perennial, 1999.

Ramón y Cajal, S. *Advice for a Young Investigator*. Translated by N Swanson and LW Swanson. Cambridge, MA: MIT Press, 1999 [1897].

———. *Recollections of My Life*. Cambridge, MA: MIT Press, 1937. Originally published as *Recuerdos de Mi Vida*, translated by EH Craigie (Madrid, 1901–1917).

Rawson, KA, and J Dunlosky. "Optimizing schedules of retrieval practice for durable and efficient learning: How much is enough?" *Journal of Experimental Psychology: General* 140, 3 (2011): 283–302.

Rivard, LP, and SB Straw. "The effect of talk and writing on learning science: An exploratory study." *Science Education* 84, 5 (2000): 566–593.

Rocke, AJ. *Image and Reality*. Chicago: University of Chicago Press, 2010.

Roediger, HL, and AC Butler. "The critical role of retrieval practice in long-term retention." *Trends in Cognitive Sciences* 15, 1 (2011): 20–27.

Roediger, HL, and JD Karpicke. "The power of testing memory: Basic research and implications for educational practice." *Perspectives on Psychological Science* 1, 3 (2006): 181–210.

Roediger, HL, and MA Pyc. "Inexpensive techniques to improve education: Applying cognitive psychology to enhance educational practice." *Journal of Applied Research in Memory and Cognition* 1, 4 (2012): 242–248.

Rohrer, D., Dedrick, R. F., & Burgess, K. (in press). The benefit of interleaved mathematics practice is not limited to superficially similar kinds of problems. *Psychonomic Bulletin & Review.*

Rohrer, D, and H Pashler. "Increasing retention without increasing study time." *Current Directions in Psychological Science* 16, 4 (2007): 183–186.

———. "Recent research on human learning challenges conventional instructional strategies." *Educational Researcher* 39, 5 (2010): 406–412.

Root-Bernstein, RS, and MM Root-Bernstein. *Sparks of Genius.* New York: Houghton Mifflin, 1999.

Ross, J, and KA Lawrence. "Some observations on memory artifice." *Psychonomic Science* 13, 2 (1968): 107–108.

Schoenfeld, AH. "Learning to think mathematically: Problem solving, metacognition, and sense-making in mathematics." In *Handbook for Research on Mathematics Teaching and Learning,* edited by D Grouws. 334–370, New York: Macmillan, 1992.

Schutz, LE. "Broad-perspective perceptual disorder of the right hemisphere." *Neuropsychology Review* 15, 1 (2005): 11–27.

Scullin, MK, and MA McDaniel. "Remembering to execute a goal: Sleep on it!" *Psychological Science* 21, 7 (2010): 1028–1035.

Shannon, BJ, et al. "Premotor functional connectivity predicts impulsivity in juvenile offenders." *Proceedings of the National Academy of Sciences* 108, 27 (2011): 11241–11245.

Shaw, CA, and JC McEachern, eds. *Toward a Theory of Neuroplasticity.* New York: Psychology Press, 2001.

Silverman, L. *Giftedness 101.* New York: Springer, 2012.

Simon, HA. "How big is a chunk?" *Science* 183, 4124 (1974): 482–488.

Simonton, DK. *Creativity in Science.* New York: Cambridge University Press, 2004.

———. *Scientific Genius.* New York: Cambridge University Press, 2009.

Sklar, AY, et al. "Reading and doing arithmetic nonconsciously." *Proceedings of the National Academy of Sciences* 109, 48 (2012): 19614–19619.

Smoker, TJ, et al. "Comparing memory for handwriting versus typing." In *Proceedings of the Human Factors and Ergonomics Society Annual Meeting,* 53 (2009): 1744–1747.

Solomon, I. "Analogical transfer and 'functional fixedness' in the science classroom." *Journal of Educational Research* 87, 6 (1994): 371–377.

Spear, LP. "Adolescent neurodevelopment." *Journal of Adolescent Health* 52, 2 (2013): S7–S13.

Steel, P. "The nature of procrastination: A meta-analytic and theoretical review of quintessential self-regulatory failure." *Psychological Bulletin* 133, 1 (2007): 65–94.

———. *The Procrastination Equation.* New York: Random House, 2010.

Stickgold, R, and JM Ellenbogen. "Quiet! Sleeping brain at work." *Scientific American Mind* 19, 4 (2008): 22–29.

Sweller, J, et al. *Cognitive Load Theory.* New York: Springer, 2011.

Takeuchi, H, et al. "The association between resting functional connectivity and creativity." *Cerebral Cortex* 22, 12 (2012): 2921–2929.

———. "Failing to deactivate: The association between brain activity during a working memory task and creativity." *NeuroImage* 55, 2 (2011): 681–687.

Taylor, K, and D Rohrer. "The effects of interleaved practice." *Applied Cognitive Psychology* 24, 6 (2010): 837–848.

Thomas, C, and CI Baker. "Teaching an adult brain new tricks: A critical review of evidence for training-dependent structural plasticity in humans." *NeuroImage* 73 (2013): 225–236.

Thompson-Schill, SL, et al. "Cognition without control: When a little frontal lobe goes a long way." *Current Directions in Psychological Science* 18, 5 (2009): 259–263.

Tice, DM, and RF Baumeister. "Longitudinal study of procrastination, performance, stress, and health: The costs and benefits of dawdling." *Psychological Science* 8, 6 (1997): 454–458.

Thurston, W. P. (1990). "Mathematical education." *Notices of the American Mathematical Society*, 37 (7), 844–850.

University of Utah Health Care Office of Public Affairs. "Researchers debunk myth of 'right-brain' and 'left-brain' personality traits." 2013. http://healthcare.utah .edu/publicaffairs/news/current/08-14-13_brain_personality_traits.html.

Van Praag, H, et al. "Running increases cell proliferation and neurogenesis in the adult mouse dentate gyrus." *Nature Neuroscience* 2, 3 (1999): 266–270.

Velay, J-L, and M Longcamp. "Handwriting versus typewriting: Behavioural and cerebral consequences in letter recognition." In *Learning to Write Effectively*, edited by M Torrance et al. Bradford, UK: Emerald Group, 2012: 371–373.

Wamsley, EJ, et al. "Dreaming of a learning task is associated with enhanced sleep-dependent memory consolidation." *Current Biology* 20, 9 (2010): 850–855.

Wan, X, et al. "The neural basis of intuitive best next-move generation in board game experts." *Science* 331, 6015 (2011): 341–346.

Weick, KE. "Small wins: Redefining the scale of social problems." *American Psychologist* 39, 1 (1984): 40–49.

White, HA, and P Shah. "Creative style and achievement in adults with attention-deficit/hyperactivity disorder." *Personality and Individual Differences* 50, 5 (2011): 673–677.

———. "Uninhibited imaginations: Creativity in adults with attention-deficit/ hyperactivity disorder." *Personality and Individual Differences* 40, 6 (2006): 1121–1131.

Wilson, T. *Redirect*. New York: Little, Brown, 2011.

Wissman, KT, et al. "How and when do students use flashcards?" *Memory* 20, 6 (2012): 568–579.

Xie, L, et al. "Sleep drives metabolite clearance from the adult brain." *Science* 342, 6156 (2013): 373–377.

credits

19. Puzzle of man in Mustang, partly assembled, image © 2014 Kevin Mendez and Philip Oakley
20. Puzzle of man in Mustang, mostly assembled, image © 2014 Kevin Mendez and Philip Oakley
21. Chunking a concept into a ribbon, image courtesy the author
22. Skipping to the right solution, image © 2014 Kevin Mendez
23. Practice makes permanent, image © 2014 Kevin Mendez
24. Puzzle of Mustang, faint and partly assembled, image © 2014 Kevin Mendez
25. Neural hook, image © 2014 Kevin Mendez
26. Paul Kruchko and family, photo courtesy Paul Kruchko
27. Procrastination funneling, image © 2014 Kevin Mendez
28. Norman Fortenberry, image © 2011, American Society for Engineering Education; photo by Lung-I Lo
29. Many tiny accomplishments, image courtesy the author
30. Pomodoro timer, Autore: Francesco Cirillo rilasciata a Erato nelle sottostanti licenze seguirá OTRS, http://en.wikipedia.org/wiki/File:Il_pomodoro.jpg
31. Physicist Antony Garrett Lisi surfing, author Cjean42, http://en.wikipedia.org/wiki/File:Garrett_Lisi_surfing.jpg
32. Oraldo "Buddy" Saucedo, photo courtesy of Oraldo "Buddy" Saucedo
33. Neel Sundaresan, photo courtesy Toby Burditt
34. Zombie task list, image © 2014 Kevin Mendez
35. Mary Cha, photo courtesy Mary Cha
36. Smiling zombie, image © 2014 Kevin Mendez
37. Photo of Joshua Foer, © Christopher Lane
38. Flying mule, image © 2014 Kevin Mendez
39. Zombie hand mnemonic, image © 2014 Kevin Mendez
40. Memory palace, image © 2014 Kevin Mendez
41. Sheryl Sorby, photo by Brockit, Inc., supplied courtesy Sheryl Sorby
42. Monkeys in a ring, from *Berichte der Durstigen Chemischen Gesellschaft* (1886), p. 3536; benzene ring, modified from http://en.wikipedia.org/wiki/File:Benzene-2D-full.svg
43. Metabolic vampires, image © 2014 Kevin Mendez
44. Jonathon Strong, photo courtesy Jonathon Strong
45. Zombie baseball player, image © 2014 Kevin Mendez
46. Nick Appleyard, photo courtesy Nick Appleyard
47. Santiago Ramón y Cajal, by kind permission of Santiago Ramón y Cajal´s heirs, with the gracious assistance of Maria Angeles Ramón y Cajal
48. Rippling neural ribbons, image courtesy author
49. Photons, illustration courtesy Marco Bellini, Instituto Nazionale di Ottica—CNR, Florence, Italy
50. Barbara McClintock, photo courtesy Smithsonian Institution Archives, image #SIA2008–5609
51. Ben Carson, photo courtesy Johns Hopkins Medicine

52. Nicholas Wade, photo courtesy Nicholas Wade
53. Ischemic stroke, CT scan of the brain with an MCA infarct, by Lucien Monfils, http://en.wikipedia.org/wiki/File:MCA_Territory_Infarct.svg
54. Niels Bohr lounging with Einstein in 1925, picture by Paul Ehrenfest, http://en.wikipedia.org/wiki/File:Niels_Bohr_Albert_Einstein_by_Ehrenfest.jpg
55. Brad Roth, photo by Yang Xia, courtesy Brad Roth
56. Richard M. Felder, courtesy Richard M. Felder
57. Sian Beilock, courtesy University of Chicago
58. Dime solution, image courtesy the author

index

Page numbers in *italics* indicate photographs or illustrations.

Gabora, Liane, 32
Galois, Évariste, 224
Gamache, Robert R., 72
Gashaj, Michael, 137
Gates, Bill, 66, 216
Gazzaniga, Michael, 228
generation (recalling) effect, chunking, 115
genetic transposition ("jumping genes"), 206
genius envy, 185–89, 190
GI Bill, 5
goals, setting, 136, 137, 141, 152
Golden Apple Award, 208
Goldman Sachs, 199
Google, 27, 127, 138
Gordon, Cassandra, 41
Granovetter, Mark, 231
Gray-Grant, Daphne, 131
group work, 120, 130, 231–33, 234, 235, 239, 240, 241, 255, 259
Gruber, Howard, 30

habits. *See* zombies
hand bones mnemonic, 176
hand writing. *See* writing by hand
hard-start–jump-to-easy technique, 241–44, 245–46, 248, 249
hard tasks, 114, 116, 122, 148–49
Hardy, G. H., 223
harnessing, extending abilities, 32–33
harnessing your zombies (habits), 84, 95–101, *97*
Hasan, Yusra, 96
Hebert, Susan Sajna, 246
hidden meanings in equations, 203–5, 211, 212
highlighting text, 62, 125, 178, 259
highly attentive state network, 11
hitting the wall (knowledge collapse), chunking, 118, 123
homework and test preparation, 240
hostage negotiation, 147

illusions of competence, 61–68, *64, 67,* 77, 79, 117, 125
"impostor phenomenon," 188
index cards example, 75–76
Inspire! program, 124
intellectual snipers caution, 219–21, 222
intention to learn and learning, 62

interleaving vs. overlearning, *74,* 74–78, 113, 173
Intermediate Physics for Medicine and Biology (Roth), 236
internalizing concepts and solutions, 6, 73
introverts and teamwork, 233
intuition problem solving, 67, *67,* 236–37, 247
intuitive understanding, 183–85, *185,* 190
inventions, enhancing, 112, 113
Iraq, 80
isolation, 126, 130, 139, 153

James, William, 119
Jeshurun, Weston, 150
jingles, 163
Jobs, Steve, 216
Johansson, Frans, 144–45
Johnson, Steven, 66
Jordan, 168
"jumping genes" (genetic transposition), 206
"just this one time" phenomenon, 135–36
juvenile delinquents, 193–94, 199

Kamkwamba, William, 215
Kanigel, Robert, 223
Karpicke, Jeffrey, xvii–xviii, 61–62
Kasparov, Garry, 9–10, *10,* 37
keeping up with the intellectual Joneses, 36
keys to becoming an "Equation Whisperer." *See* chunking
"keystone" bad habit, procrastination, 86
knowledge collapse (hitting the wall), chunking, 118, 123
knowledge vs. memory trick, 176
Koehler, William, 180–81
Kruchko, Paul, *80,* 80–82

labels and confidence, 192
language-learning skills, 1, 4, 5, 6, 14, 16, 53, 63, 70, 118, 145, 198, 209, 210
Law of Serendipity, ix, 66, 116, 122, 137, 256
learned industriousness, 99
learning more effectively, 6–7
   See also math and science, learning
learning on your own, 213–16, *214,* 218, 221, 222
learning to appreciate your talent. *See* appreciating your talent